An Introduction to Environmental Biophysics

Second Edition

Gaylon S. Campbell
John M. Norman

An Introduction to Environmental Biophysics

Second Edition

With 81 Illustrations

 Springer

Gaylon S. Campbell
Decagon Devices, Inc.
950 NE Nelson Ct.
Pullman, WA 99163
USA

John M. Norman
University of Wisconsin
College of Agricultural and
  Life Sciences Soils
Madison, WI 53705
USA

Library of Congress Cataloging-in-Publication Data
Campbell, Gaylon S.
    Introduction to environmental biophysics/G. S. Campbell, J. M.
Norman. -- 2nd ed.
      p. cm.
    Includes bibliographical references and index.
    ISBN 0-387-94937-2 (softcover)
    1. Biophysics. 2. Ecology. I. Norman, John M. II. Title.
CH505.C34   1998
571.4--dc21   97-15706

ISBN 0-387-94937-2
Printed on acid-free paper

Printed in the United States of America.   (EB)

9 8 7 6 5 4

springeronline.com

Preface to the Second Edition

The objectives of the first edition of "An Introduction to Environmental Biophysics" were "to describe the physical microenvironment in which living organisms reside" and "to present a simplified discussion of heat and mass transfer models and apply them to exchange processes between organisms and their surroundings." These remain the objectives of this edition. This book is used as a text in courses taught at Washington State University and University of Wisconsin and the new edition incorporates knowledge gained through teaching this subject over the past 20 years. Suggestions of colleagues and students have been incorporated, and all of the material has been revised to reflect changes and trends in the science.

Those familiar with the first edition will note that the order of presentation is changed somewhat. We now start by describing the physical environment of living organisms (temperature, moisture, wind) and then consider the physics of heat and mass transport between organisms and their surroundings. Radiative transport is treated later in this edition, and is covered in two chapters, rather than one, as in the first edition. Since remote sensing is playing an increasingly important role in environmental biophysics, we have included material on this important topic as well. As with the first edition, the final chapters are applications of previously described principles to animal and plant systems.

Many of the students who take our courses come from the biological sciences where mathematical skills are often less developed than in physics and engineering. Our approach, which starts with more descriptive topics, and progresses to topics that are more mathematically demanding, appears to meet the needs of students with this type of background. Since we expect students to develop the mathematical skills necessary to solve problems in mass and energy exchange, we have added many example problems, and have also provided additional problems for students to work at the end of chapters.

One convention the reader will encounter early in the book, which is a significant departure from the first edition, is the use of molar units for mass concentrations, conductances, and fluxes. We have chosen this unit convention for several reasons. We believe molar units to be fundamental, so equations are simpler with fewer coefficients when molar units are used. Also, molar units are becoming widely accepted in biological science disciplines for excellent scientific reasons (e.g., photosynthetic light reactions clearly are driven by photons of light and molar units are required to describe this process.) A coherent view of the connectedness of biological organisms and their environment is facilitated by a uniform system of units. A third reason for using molar units comes from the fact that, when diffusive conductances are expressed in molar units, the numerical values are virtually independent of temperature and pressure. Temperature and pressure effects are large enough in the old system to require adjustments for changes in temperature and pressure. These temperature and pressure effects were not explicitly acknowledged in the first edition, making that approach look simpler; but students who delved more deeply into the problem found that, to do the calculations correctly, a lot of additional work was required. A fourth consideration is that use of a molar unit immediately raises the question "moles of what?" The dependence of the numerical value of conductance on the quantity that is diffusing is more obvious than when units of m/s are used. This helps students to avoid using a diffusive conductance for water vapor when estimating a flux of carbon dioxide, which would result in a 60 percent error in the calculation. We have found that students adapt readily to the consistent use of molar units because of the simpler equations and explicit dependencies on environmental factors. The only disadvantage to using molar units is the temporary effort required by those familiar with other units to become familiar with "typical values" in molar units.

A second convention in this book that is somewhat different from the first edition is the predominant use of conductance rather that resistance. Whether one uses resistance or conductance is a matter of preference, but predominant use of one throughout a book is desirable to avoid confusion. We chose conductance because it is directly proportional to flux, which aids in the development of an intuitive understanding of transport processes in complex systems such as plant canopies. This avoids some confusion, such as the common error of averaging leaf resistances to obtain a canopy resistance. Resistances are discussed and occasionally used, but generally to avoid unnecessarily complicated equations in special cases.

A third convention that is different from the first edition is the use of surface area instead of "projected area." This first appears in the discussion of the leaf energy budget and the use of "view factors." Because many biophysicists work only with flat leaves, the energy exchange equations for leaves usually are expressed in terms of the "one-sided" leaf area; this is the usual way to characterize the area of flat objects. If the energy balance is generalized to nonflat objects, such as animal bodies or appendages, tree trunks or branches, or conifer needles, then this "one-side" area is subject to various interpretations and serious confusion can result. Errors of a factor of two frequently occur and the most experienced biophysicist has encountered difficulty at one time or another with this problem. We believe that using element surface area and radiation "view factors" are the best way to resolve this problem so that misinterpretations do not occur. For those interested only in exchanges with flat leaves, the development in this book may seem somewhat more complicated. However, "flat leaf" versions of the equations are easy to write and when interest extends to nonflat objects this analysis will be fully appreciated. When extending energy budgets to canopies we suggest hemi-surface area, which is one-half the surface area. For canopies of flat leaves, the hemi-surface area index is identical to the traditional leaf area index; however for canopies of nonflat leaves, such as conifer needles, the hemi-surface area index is unambiguous while "projected" leaf area index depends on many factors that often are not adequately described.

One convention that remains the same as the first edition is the use of J/kg for water potential. Although pressure units (kPa or MPa) have become popular in the plant sciences, potential is an energy per unit mass and the J/kg unit is more fundamental and preferred. Fortunately, J/kg and kPa have the same numerical value so conversions are simple.

As with the previous edition, many people contributed substantially to this book. Students in our classes, as well as colleagues, suggested better ways of presenting material. Several publishers gave permission to use previously published materials. Marcello Donatelli checked the manuscript for errors and prepared the manuscript and figures to be sent to the publisher. The staff at Springer-Verlag were patient and supportive through the inevitable delays that come with full schedules. We are also grateful to our wives and families for their help and encouragement in finishing this project. Finally, we would like to acknowledge the contributions of the late Champ B. Tanner. Most of the material in this book was taught and worked on in some form by Champ during his years of teaching and research at University of Wisconsin. Both of us have been deeply influenced by his teaching and his example. We dedicate this edition to him.

G. S. Campbell
J. M. Norman
Pullman and Madison, 1997

Preface to the
First Edition

The study of environmental biophysics probably began earlier in man's history than that of any other science. The study of organism—environment interaction provided a key to survival and progress. Systematic study of the science and recording of experimental results goes back many hundreds of years. Benjamin Franklin, the early American statesmen, inventor, printer, and scientist studied conduction, evaporation, and radiation. One of his observation is as follows:

> My desk on which I now write, and the lock of my desk, are both exposed to the same temperature of the air, and have therefore the same degree of heat or cold; yet if I lay my hand successively on the wood and on the metal, the latter feels much the coldest, not that it is really so, but being a better conductor, it more readily than the wood takes away and draws into itself the fire that was in my skin.[1]

Progress in environmental biophysics, since the observation of Franklin and others, has been mainly in two areas: use of mathematical models to quantify rates of heat and mass transfer and use of the continuity equation that has led to energy budget analyses. In quantification of heat- and mass-transfer rates, environmental biophysicists have followed the lead of physics and engineering. There, theoretical and empirical models have been derived that can be applied to many of the transport problems encountered by the design engineer. The same models were applied to transport processes between living organisms and their surroundings.

This book is written with two objectives in mind. The first is to describe the physical micro environment in which living organisms reside. The second is to present a simplified discussion of heat- and mass-transfer models and apply them to exchange processes between organisms and their surroundings. One might consider this a sort of engineering approach to environmental biology, since the intent to teach the student to calculate actual transfer rates, rather than just study the principles involved.

---

[1] From a letter to John Lining, written April 14, 1757. The entire letter, along with other scientific writings by Franklin, can be found in Reference [1.2].

Numerical examples are presented to illustrate many of the principles, and are given at the end of each chapter to help the student develop skills using the equations. Working of problems should be considered as essential to gaining an understanding of modern environmental biophysics as it is to any course in physics or engineering. The last four chapters of the book attempt to apply physical principles to exchange processes of living organisms, the intent was to indicate approaches that either could be or have been used to solve particular problems. The presentation was not intended to be exhaustive, and in many cases, assumptions made will severely limit the applicability of the solutions. It is hoped that the reader will find these examples helpful but will use the principles presented in the first part of the book to develop his own approaches to problems, using assumptions that fit the particular problem of interest.

Literature citation have been given at the end of each chapter to indicate sources of additional material and possibilities for further reading. Again, the citations were not meant to be exhaustive.

Many people contributed substantially to this book. I first became interested in environmental biophysics while working as an undergraduate in the laboratory of the late Sterling Taylor. Walter Gardner has contributed substantially to my understanding of the subject through comments and discussion, and provided editorial assistance on early chapters of the book. Marcel Fuchs taught me about light penetration in plant canopies, provided much helpful discussion on other aspects of the book, and read and commented on the entire manuscript. James King read Chapters 7 and 8 and made useful criticisms which helped the presentation. He and his students in zoology have been most helpful in providing discussion and questions which led to much of the material presented in Chapter 7. Students in my Environmental Biophysics classes have offered many helpful criticisms to make the presentation less ambiguous and, I hope, more understandable. Several authors and publishers gave permission to use figures, Karen Ricketts typed all versions of the manuscript, and my wife, Judy, edited the entire manuscript and offered the help and encouragement necessary to bring this project to completion. To all of these people, I am most grateful.

Pullman, 1977   G. S. C.

Contents

List of Symbols

A	$\{\text{mol m}^{-2}\text{ s}^{-1}\}$	*carbon assimilation rate*
$A(0)$	$\{\text{C}\}$	*amplitude of the diurnal soil surface temperature*
A_w		*plant available water*
B	$\{\text{W/m}^2\}$	*flux density of blackbody radiation*
c	$\{\text{m/s}\}$	*speed of light*
c		*fraction of sky covered with cloud*
c_p	$\{\text{J mol}^{-1}\text{ C}^{-1}\}$	*specific heat of air at constant pressure*
c_s	$\{\text{J kg}^{-1}\text{ C}^{-1}\}$	*specific heat of soil*
C_j	$\{\text{mol/mol}\}$	*concentration of gas j in air*
C	$\{\text{mol/kg}\}$	*concentration of solute in osmotic solution*
d	$\{\text{m}\}$	*zero plane displacement*
d	$\{\text{m}\}$	*characteristic dimension*
D	$\{\text{m}\}$	*soil damping depth*
D	$\{\text{kPa}\}$	*vapor deficit of air*
D_H	$\{\text{m}^2\text{ /s}\}$	*thermal diffusivity*
e	$\{\text{J}\}$	*energy of one photon*
e	$\{\text{kg/MJ}\}$	*radiation conversion efficiency for crops*
e	$\{\text{kPa}\}$	*vapor pressure of water*
e_a	$\{\text{kPa}\}$	*partial pressure of water vapor in air*
$e_s(T)$	$\{\text{kPa}\}$	*saturation vapor pressure of water at temperature T*
E	$\{\text{mol m}^{-2}\text{ s}^{-1}\}$	*evaporation rate for water*
E_r	$\{\text{mol m}^{-2}\text{ s}^{-1}\}$	*respiratory evaporative water loss*
E_s	$\{\text{mol m}^{-2}\text{ s}^{-1}\}$	*skin evaporative water loss*
f		*fraction of radiation intercepted by a crop canopy*
f_{ds}		*fraction of downscattered radiation in a particular waveband*
F_a		*view factor for atmospheric thermal radiation*
F_d		*view factor for diffuse solar radiation*
F_g		*view factor for ground thermal radiation*
F_p		*view factor for solar beam*

$F_j(z)$	$\{\text{mol m}^{-2}\,\text{s}^{-1}\}$	*flux density of j at location z*
g	$\{\text{m/s}^2\}$	*gravitational constant*
g_H	$\{\text{mol m}^{-2}\,\text{s}^{-1}\}$	*conductance for heat*
g_{Ha}	$\{\text{mol m}^{-2}\,\text{s}^{-1}\}$	*boundary layer conductance for heat*
g_{Hb}	$\{\text{mol m}^{-2}\,\text{s}^{-1}\}$	*whole body conductance (coat and tissue) for an animal*
g_{Hc}	$\{\text{mol m}^{-2}\,\text{s}^{-1}\}$	*coat conductance for heat*
g_{Hr}	$\{\text{mol m}^{-2}\,\text{s}^{-1}\}$	*sum of boundary layer and radiative conductances*
g_{Ht}	$\{\text{mol m}^{-2}\,\text{s}^{-1}\}$	*tissue conductance for heat*
g_r	$\{\text{mol m}^{-2}\,\text{s}^{-1}\}$	*radiative conductance*
g_v	$\{\text{mol m}^{-2}\,\text{s}^{-1}\}$	*conductance for vapor*
g_{va}	$\{\text{mol m}^{-2}\,\text{s}^{-1}\}$	*boundary layer conductance for vapor*
g_{vs}	$\{\text{mol m}^{-2}\,\text{s}^{-1}\}$	*surface or stomatal conductance for vapor*
G	$\{\text{W/m}^2\}$	*soil heat flux density*
h	$\{\text{m}\}$	*canopy height*
h	$\{\text{J s}\}$	*Planck's constant*
h_r		*relative humidity*
H	$\{\text{W/m}^2\}$	*sensible heat flux density*
J_w	$\{\text{kg m}^{-2}\,\text{s}^{-1}\}$	*water flux density*
k	$\{\text{W m}^{-1}\,\text{C}^{-1}\}$	*thermal conductivity*
k	$\{\text{J/K}\}$	*Boltzmann constant*
K		*canopy extinction coefficient*
$K_{be}(\psi)$		*extinction coefficient of a canopy of black leaves with an ellipsoidal leaf angle distribution for beam radiation*
K_d		*extinction coefficient of a canopy of black leaves for diffuse radiation*
K_m	$\{\text{m}^2\,\text{/s}\}$	*eddy diffusivity for momentum*
K_H	$\{\text{m}^2\,\text{/s}\}$	*eddy diffusivity for heat*
K_v	$\{\text{m}^2\,\text{/s}\}$	*eddy diffusivity for vapor*
K_s	$\{\text{kg s m}^{-3}\}$	*saturated hydraulic conductivity of soil*
L	$\{\text{m}^2\,\text{/m}^2\}$	*total leaf area index of plant canopy*
L_{oe}	$\{\text{W/m}^2\}$	*emitted long-wave radiation*
L_t		*leaf area index above some height in a canopy*
L_t^*		*sunlit leaf area index in a complete canopy*
m		*airmass number*
M	$\{\text{W/m}^2\}$	*metabolic rate*
M_b	$\{\text{w/m}^2\}$	*basal metabolic rate*
M_j	$\{\text{g/mol}\}$	*molar mass of gas j*
n_j	$\{\text{mol}\}$	*number of moles of gas j*
p_j	$\{\text{kPa}\}$	*partial pressure of gas j*
p_a	$\{\text{kPa}\}$	*atmospheric pressure*
q	$\{\text{g/g}\}$	*specific humidity (mass of water vapor divided by mass of moist air)*
Q_p	$\{\mu\text{mol m}^{-2}\,\text{s}^{-1}\}$	*PAR photon flux density*

r_H	$\{m^2 \text{ s mol}^{-1}\}$	heat transfer resistance $(1/g_H)$
r_v	$\{m^2 \text{ s mol}^{-1}\}$	vapor transfer resistance $(1/g_v)$
R	$\{J \text{ mol}^{-1} C^{-1}\}$	gas constant
R_{abs}	$\{W/m^2\}$	absorbed short- and long-wave radiation
R_d	$\{\mu mol \text{ m}^{-2} \text{ s}^{-1}\}$	dark respiration rate of leaf
R_L	$\{m^4 \text{ s}^{-1} \text{ kg}^{-1}\}$	resistance to water flow through a plant leaf
R_n	$\{W/m^2\}$	net radiation
R_R	$\{m^4 \text{ s}^{-1} \text{ kg}^{-1}\}$	resistance to water flow through a plant root
s	$\{C^{-1}\}$	slope of saturation mole fraction function (Δ/p_a)
S_b	$\{W/m^2\}$	flux density of solar radiation on a horizontal surface
S_d	$\{W/m^2\}$	flux density of diffuse radiation on a surface
S_p	$\{W/m^2\}$	flux density of solar radiation perpendicular to the solar beam
S_r	$\{W/m^2\}$	flux density of reflected solar radiation
S_{po}	$\{W/m^2\}$	the solar constant
S_t	$\{W/m^2\}$	flux density of total solar radiation
t	$\{s\}$	time
t_o	$\{hr\}$	time of solar noon
$T(z)$	$\{C\}$	temperature at height z
$T(t)$	$\{C\}$	temperature at time t
T_d	$\{C\}$	dew point temperature
T_e	$\{C\}$	operative temperature
T_{es}	$\{C\}$	standard operative temperature
T_{eh}	$\{C\}$	humid operative temperature
T_o	$\{C\}$	apparent aerodynamic surface temperature
T_{ave}	$\{C\}$	average soil temperature
T_b	$\{C\}$	base temperature for biological development
T_{xi}	$\{C\}$	maximum temperature on day i
T_{ni}	$\{C\}$	minimum temperature on day i
$\mathbf{T}$	$\{K\}$	kelvin temperature
u^*	$\{m/s\}$	friction velocity of wind
V_m	$\{\mu mol \text{ m}^{-2} \text{ s}^{-2}\}$	maximum Rubisco capacity per unit leaf area
w	$\{g/g\}$	mixing ratio (mass of water vapor divided by mass of dry air)
w	$\{g/g\}$	mass wetness of soil
x		average area of canopy elements projected on to the horizontal plane divided by the average area projected on to a vertical plane

z	{m}	height in atmosphere or depth in soil
z_H	{m}	roughness length for heat
z_M	{m}	roughness length for momentum

Greek

α		absorptivity for radiation
α_s		absorptivity for solar radiation
α_L		absorptivity for longwave radiation
β	{degrees}	solar elevation angle
δ	{degrees}	solar declination
Δ	{kPa/C}	slope of the saturation vapor pressure function
ϵ		emissivity
ϵ_{ac}		emissivity of clear sky
$\epsilon_a(c)$		emissivity of sky with cloudiness c
ϵ_s		emissivity of surface
γ	{C^{-1}}	thermodynamic psychrometer constant (c_p/λ)
γ^*	{C^{-1}}	apparent psychrometer constant
Γ^*	{mol/mol}	light compensation point
$\Gamma(t)$		dimensionless diurnal function for estimating hourly air temperature
ϕ		osmotic coefficient
ϕ	{degrees}	latitude
ϕ_M		diabatic influence factor for momentum
ϕ_H		diabatic influence factor for heat
ϕ_v		diabatic influence factor for vapor
Φ	{W/m^2}	flux density of radiation
κ	{m^2/s}	soil thermal diffusivity
λ	{J/mol}	latent heat of vaporization of water
λ	{μ m}	wavelength of electromagnetic radiation
ψ	{J/kg}	water potential
ψ	{degrees}	solar zenith angle
Ψ_m		diabatic correction for momentum
Ψ_H		diabatic correction for heat
$\hat{\rho}$	{mol m^{-3}}	molar density of air
ρ		leaf reflectivity
ρ_b	{kg/m^3}	bulk density of soil
$\rho_{b,cpy}^H$		bihemispherical reflectance of a canopy of horizontal leaves with infinte LAI
$\rho_{b,cpy}^*$		canopy bihemispherical reflectance for diffuse radiation and a canopy of infinite LAI
$\rho_{b,cpy}^*(\psi)$		canopy directional-hemisperical reflectance for beam radiation incident at angle Ψ for a canopy of infinite LAI
ρ_j	{g/m^{-3}}	density of gas j in air

θ	{degrees}	angle between incident radiation and a normal to a surface
θ	{$m^3\ m^{-3}$}	volume wetness of soil
τ	{day deg}	thermal time
τ	{s}	period of periodic temperature variations
τ		sky transmittance
τ	{s}	thermal time constant of an animal
τ_b		fraction of beam radiation transmitted by a canopy
$\tau_b(\psi)$		fraction of beam radiation that passes through a canopy without being intercepted by any objects
$\tau_{bt}(\psi)$		fraction of incident beam radiation transmitted by a canopy including scattered and unintercepted beam radiation
τ_d		fraction of diffuse radiation transmitted by a canopy
ζ		atmospheric stability parameter
ω	{s^{-1}}	angular frequency of periodic temperature variations

Introduction

1

The discipline of environmental biophysics relates to the study of energy and mass exchange between living organisms and their environment. The study of environmental biophysics probably began earlier than that of any other science, since knowledge of organism–environment interaction provided a key to survival and progress. Systematic study of the science and recording of experimental results, however, goes back only a few hundred years. Recognition of environmental biophysics as a discipline has occurred just within the past few decades.

Recent progress in environmental biophysics has been mainly in two areas: use of mathematical models to quantify rates of energy and mass transfer and use of conservation principles to analyze mass and energy budgets of living organisms. In quantification of energy and mass transfer rates, environmental biophysicists have followed the lead of classical physics and engineering. There, theoretical and empirical models have been derived that can be applied to many of the transport problems encountered by the design engineer. These same models can be applied to transport processes between living organisms and their surroundings.

This book is written with two objectives in mind. The first is to describe and model the physical microenvironment in which living organisms reside. The second is to present simple models of energy and mass exchange between organisms and their microenvironment with models of organism response to these fluxes of energy and matter. One might consider this a combined science and engineering approach to environmental biology because the intent is to teach the student to calculate actual transfer rates and to understand the principles involved. Numerical examples are presented to illustrate many of the principles, and problems are given at the end of each chapter to help the student develop skill in using the equations. Working the problems should be considered as essential to gaining an understanding of modern environmental biophysics as it is to any course in physics or engineering.

A list of symbols with definitions is provided at the beginning of this book, and tables of data and conversions are in appendices at the end of the book. It would be a good idea to look at those now, and use them frequently as you go through the book. References are given at the end of each chapter to indicate sources of the materials presented and to provide additional information on subjects that can be treated only briefly in the text. Citations certainly are not intended to be exhaustive, but should lead serious students into the literature.

The effects of the physical environment on behavior and life are such an intimate part of our everyday experience that one may wonder at the need to study them. Heat, cold, wind, and humidity have long been common terms in our language, and we may feel quite comfortable with them. However, we often misinterpret our interaction with our environment and misunderstand the environmental variables themselves. Benjamin Franklin, the early American statesman, inventor, printer, and scientist alludes to the potential for misunderstanding these interactions. In a letter to John Lining, written April 14, 1757 he wrote (Seeger, 1973):

My desk on which I now write, and the lock of my desk, are both exposed to the same temperature of the air, and have therefore the same degree of heat or cold; yet if I lay my hand successively on the wood and on the metal, the latter feels much the coldest, not that it is really so, but being a better conductor, it more readily than the wood takes away and draws into itself the fire that was in my skin.

Franklin's experiment and the analysis he presents help us understand that we do not sense temperature; we sense changes in temperature which are closely related to the flow of heat toward or away from us. The heat flux, or rate of heat flow depends on a temperature difference, but it also depends on the resistance or conductance of the intervening medium.

Careful consideration will indicate that essentially every interaction we have with our surroundings involves energy or mass exchange. Sight is possible because emitted or reflected photons from our surroundings enter the eye and cause photochemical reactions at the retina. Hearing results from the absorption of acoustic energy from our surroundings. Smell involves the flux of gases and aerosols to the olfactory sensors. Numerous other sensations could be listed such as sunburn, heat stress, cold stress, and each involves the flux of something to or from the organism. The steady-state exchange of most forms of matter and energy can be expressed between organisms and their surroundings as:

$$\text{Flux} = g(C_s - C_a) \qquad (1.1)$$

where C_s is the concentration at the organism exchange surface, C_a is the ambient concentration, and g is an exchange conductance. As already noted, our senses respond to fluxes but we interpret them in terms of ambient concentrations. Even if the concentration at the organism were constant (generally not the case) our judgment about ambient concentration would always be influenced by the magnitude of the exchange conductance. Franklin's experiment illustrates this nicely. The higher conductance of the metal made it feel colder, even though the wood and the metal were at the same temperature.

1.1 Microenvironments

Microenvironments are an intimate part of our everyday life, but we seldom stop to think of them. Our homes, our beds, our cars, the sheltered side of a building, the shade of a tree, an animal's burrow are all examples of microenvironments. The "weather" in these places cannot usually be described by measured and reported weather data. The air temperature may be 10° C and the wind 5 m/s, but an insect, sitting in an animal track sheltered from the wind and exposed to solar radiation may be at a comfortable 25° C. It is the microenvironment that is important when considering organism energy exchange, but descriptions of microclimate are often complicated because the organism influences its microclimate and because microclimates are extremely variable over short distances. Specialized instruments are necessary to measure relevant environmental variables. Variables of concern may be temperature, atmospheric moisture, radiant energy flux density, wind, oxygen and CO_2 concentration, temperature and thermal conductivity of the substrate (floor, ground, etc.), and possibly spectral distribution of radiation. Other microenvironmental variables may be measured for special studies.

We first concern ourselves with a study of the environmental variables—namely, temperature, humidity, wind, and radiation. We then discuss energy and mass exchange, the fundamental link between organisms and their surroundings. Next we apply the principles of energy and mass exchange to a few selected problems in plant, animal, and human environmental biophysics. Finally, we consider some problems in radiation, heat, and water vapor exchange for vegetated surfaces such as crops or forests.

1.2 Energy Exchange

The fundamental interaction of biophysical ecology is energy exchange. Energy may be exchanged as stored chemical energy, heat energy, radiant energy, or mechanical energy. Our attention will be focused primarily on the transport of heat and radiation.

Four modes of energy transfer are generally recognized in our common language when we talk of the "hot" sun (radiative exchange) or the "cold" floor tile (conduction), the "chilling" wind (convection), or the "stifling" humidity (reduced latent heat loss). An understanding of the principles behind each of these processes will provide the background needed to determine the physical suitability of a given environment for a particular organism.

The total heat content of a substance is proportional to the total random kinetic energy of its molecules. Heat can flow from one substance to another if the average kinetic energies of the molecules in the two substances are different. Temperature is a measure of the average random kinetic energy of the molecules in a substance. If two substances at different temperatures are in contact with each other, heat is transferred from the high-temperature substance to the low by conduction, a direct molecular interaction. If you touch a hot stove, your hand is heated by conduction.

Heat transport by a moving fluid is called convection. The heat is first transferred to the fluid by conduction; the bulk fluid motion carries away the heat stored in the fluid. Most home heating systems rely on convection to heat the air and walls of the house.

Unlike convection and conduction, radiative exchange requires no intervening molecules to transfer energy from one surface to another. A surface radiates energy at a rate proportional to the fourth power of its absolute temperature. Both the sun and the earth emit radiation, but because the sun is at a higher temperature the emitted radiant flux density is much higher for the surface of the sun than for the surface of the earth. Much of the heat you receive from a campfire or a stove may be by radiation and your comfort in a room is often more dependent on the amount of radiation you receive from the walls than on the air temperature.

To change from a liquid to a gaseous state at $20°$ C, water must absorb about 2450 joules per gram (the latent heat of vaporization), almost 600 times the energy required to raise the temperature of one gram of water by one degree. Evaporation of water from an organism, which involves the latent heat required to convert the liquid water to vapor and convection of this vapor away from the organism, can therefore be a very effective mode of energy transfer. Almost everyone has had the experience of stepping out of a swimming pool on a hot day and feeling quite cold until the water dries from their skin.

1.3 Mass and Momentum Transport

Organisms in natural environments are subject to forces of wind or water and rely on mass transport to exchange oxygen and carbon dioxide. The force of wind or water on an organism is a manifestation of the transport of momentum from the fluid to the organism. Transport of momentum, oxygen, and carbon dioxide in fluids follow principles similar to those developed for convective heat transfer. Therefore, just one set of principles can be learned and applied to all three areas.

1.4 Conservation of Energy and Mass

One of the most powerful laws used in analyzing organism–environment interaction is the conservation law. It states that neither mass nor energy can be created or destroyed by any ordinary means. The application of this law is similar to the reconciliation of your checking account. You compute the deposits and withdrawals, and the difference is the balance or storage. As an example, consider the energy balance of a vegetated surface. We can write an equation representing the inputs, losses, and storage of energy as:

$$R_n + M - H - \lambda E = G \tag{1.2}$$

Here, R_n represents the net flux density of radiation absorbed by the surface, M represents the supply of energy to the surface by metabolism or absorption of energy by photosynthesis, H is the rate of loss of sensible heat (heat flow by convection or conduction due to a temperature difference), λE is the rate of latent heat loss from the surface (E is the rate of evaporation of water and λ is the latent heat of evaporation or the heat absorbed when a gram of water evaporates), and G is the rate of heat storage in the vegetation and soil. A similar equation could be written for the water balance of a vegetated surface. Since conservation laws cannot be violated, they provide valuable information about the fluxes or storage of energy or mass. In a typical application of Eq. (1.2) we might measure or estimate R_n, M, H, and G, and use the equation to compute E. Another typical application is based on the fact that R_n, H, E, and G all depend on the temperature of the surface. For some set of environmental conditions (air temperature, solar radiation, vapor pressure) there exists only one surface temperature that will balance Eq. (1.2). We use the energy budget to find that temperature.

1.5 Continuity in the Biosphere

The biosphere, which is where plants and animals live within the soil and atmospheric environments, can be thought of as a continuum of spatial scales and system components. A continuum of gas (air, water vapor, carbon dioxide, oxygen, etc.) exists from the free atmosphere to the air spaces within the soil and even the air spaces within leaves. A continuum of liquid water exists from pores within a wet soil to cells within a plant root or leaf. Throughout the system the interfaces between liquid and gas phases are the regions where water molecules go from one state to another, and these regions are where latent heat exchanges will occur. These latent heat exchanges provide a coupling between mass exchanges of water and energy exchanges. The soil is obviously linked to the atmosphere by conduction and diffusion through pores, but it is also linked to the atmosphere through the plant vascular system.

Energy and mass conservation principles can be applied to this entire system or to specific components such as a single plant, leaf, xylem vessel, or even a single cell. The transport equations can also be applied to the entire system or to a single component. Clearly, one must define carefully what portion of the system is of interest in a particular analysis.

Animals may be components of this system from microscopic organisms in films of water in the soil to larger fauna such as worms, or animals on leaves such as mites or grasshoppers, or yet larger animals in the canopy space. The particular microenvironment that the animal is exposed to will depend on interactions among components of this continuum. Animals, in turn, may alter components of the continuum; for example, herbivores that eat leaves, mites that alter stomatal function, or a disease that inhibits photosynthesis.

Energy or mass from one part or scale of this system can flow continuously into another part or scale and the consequence of this interaction is what is studied in "environmental biophysics." Water is pervasive throughout the biosphere, existing in solid, liquid, or gas states, and able to move from one place or state to another. Living organisms depend on water and have adapted in remarkable ways to its characteristics. Consider, for a moment, the flow of water in the soil–plant–atmosphere system. Rainfall impinges on the surface of the soil, after condensing from the vapor in the air, and infiltrates through the pores in response to water potential gradients to distribute water throughout the bulk soil. Water then moves through the soil, into the root, through the vascular system of a plant and into the leaf under the influence of a continuously decreasing water potential. At the leaf, liquid water is changed to water vapor, which requires a considerable amount of latent heat, and the water vapor moves in response to vapor pressure differences between the leaf and the atmosphere rather then water potential gradients. This water vapor diffuses through the stomatal pore and still-air boundary layer near the leaf surface and is carried by turbulent convection through the canopy space, the planetary boundary layer, and ultimately to the free atmosphere to be distributed around the globe and condensed again as rain. The energy required to change the liquid water in leaves to water vapor, which may be extracted from the air or provided by radiant energy from the sun, couples energy exchange to water exchange. The transport laws can be used in conjunction with conservation of mass and energy to describe the movement of water throughout this system. Even though the driving forces for movement of water may vary for different parts of the system, appropriate conductances can be defined to describe transport throughout the system. In some cases the form of the transport equation may vary for different parts of the system, but the conservation of mass principle is used to link transport equations for these various parts of the system together.

Clearly, the biosphere is a complex continuum, not only in terms of the reality of the interconnectedness of living things and their environments, but also in terms of the mathematical and physical formulations that biophysicists use to describe this remarkable system. Rational exploration of the biosphere is just beginning and it is our hope that this new "head" knowledge will be woven into your being in such a way that you will have an increased awareness of your dependence on and implicit faith in that which is not known, as well as having some simple quantitative tools at your disposal to enhance a harmonious relationship between yourself and your environment and serve others at the same time.

A schematic representation of the connectivity of energy and mass in the biosphere is illustrated in Fig. 1.1.

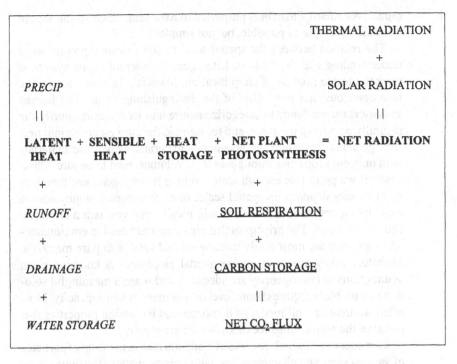

FIGURE 1.1. Schematic representation of the inter-connectedness of water (in *italics*), carbon (underlined), radiation (normal font) and energy (**bold**) budgets in a biosphere.

1.6 Models, Heterogeneity, and Scale

Throughout this book we refer to models. A model is a simple representation of a more complex form or phenomena. The term "model" is general and no interpretation of data is possible without resort to some kind of model; whether implied or explicitly declared. Many kinds of models exist and we will emphasize deterministic, mathematical models of physical and biological systems with some considerations of probability formulations. The description of natural phenomena can vary along a continuum of complexity from the trivial to the incomprehensible, and the appropriate level of complexity depends on the purpose. The application of fundamental principles to natural phenomena frequently requires adaptation of those principles or creative simplification of the natural system so that it reasonably conforms to the requirements of the underlying principles. Creative simplification of natural materials or phenomena is the "art" of environmental biophysics, and its practice depends on one's understanding of relevant fundamentals; a purpose of this book. Clearly, questions can be posed that require solutions of staggering complexity. All of nature is exceedingly complex, perhaps infinitely complex; however, insight can be gained into its complexity through the simplicity of a model. As Albert Einstein is purported to have said: "Everything should be made as simple as possible, but not simpler."

The relation between the spatial scale of some desired prediction or understanding and the scale of heterogeneity inherent in the system is essential to the process of simplification. Materials in nature tend to be heterogeneous, not pure. One of the distinguishing features of human activity is the tendency to categorize nature into its elements, purify the naturally occurring mixtures, and reassemble the pure elements into new arrangements. In nature, homogeneous materials, which are materials with uniform properties throughout their volumes, tend to be rare. Obviously, if we go to fine enough scale, nothing is homogeneous; therefore homogeneity depends on spatial scale. In environmental biophysics we consider natural materials such as soil, rock layers, vegetation mixtures, and animal coats. The principles that are commonly used in environmental biophysics are most easily understood and used with pure materials. Therefore a key aspect of environmental biophysics is knowing when assumptions of homogeneity are adequate, and when a meaningful solution to a problem requires some level of treatment of heterogeneity. Most often we treat natural media as homogeneous but assign properties that preserve the major influence of known heterogeneity.

Consider a soil, which consists of a mineral matrix made up of particles of various sizes and characteristics, with organic matter at various stages of decomposition, air, water, plant roots, worms, insects, fungi, bacteria, etc. Soil certainly is a heterogeneous medium. However, we can simulate heat transport on the scale of meters quite well by assuming soil to be homogeneous with a thermal conductivity that depends on water content, particle type and size distribution, and density. In the case of soil, the heterogeneity usually is small (millimeters) compared to the scale on which we desire to predict heat flow (meters). However, if we wish to predict the temperature and moisture environments beneath individual rocks on the surface of the soil because that is where some organism lives, then we have to deal with the apparent heterogeneity by using more complex descriptions. In the case of this heterogeneous material called "soil," various bulk properties are defined such as bulk density, heat capacity, air permeability, capillary conductivity, etc.

A second heterogeneous natural system of interest to us is a plant canopy, which consists of leaves, branches, stems, fruits, and flowers all displayed with elegance throughout some volume and able to move in response to wind, heliotropism, growth, or water stress. Simple equations have been used quite successfully to describe light penetration and canopy photosynthesis by assuming the canopy to behave like a homogeneous green slime. In spite of the seeming inappropriateness of describing photosynthesis of a 50 m tall forest canopy by radiation penetration through a green slime, a convincing intuitive argument can be forged using geometry and statistics of random distributions that is supported by direct field measurements. In fact, statistics is one of the means used to appropriately average over heterogeneity to define properties of a representative homogeneous substitute.

1.7 Applications

From the examples already given, it is quite obvious that environmental biophysics can be applied to a broad spectrum of problems. Fairly complete evaluations already exist for some problems, though much work remains to be done. Analysis of human comfort and survival in hot and cold climates requires a good understanding of the principles we will discuss. Preferred climates, survival, and food requirements of domestic and wild animals can also be considered. Plant adaptations in natural systems can be understood, and optimum plant types and growing conditions in agriculture and forestry can be selected through proper application of these principles. Even the successful architectural design of a building, which makes maximum use of solar heat and takes into account wind and other climatological variables, requires an understanding of this subject. Finally, models that forecast the weather or predict changes in past and future climates rely heavily on the principles of environmental biophysics to accommodate exchanges between the surface of the earth and the atmosphere.

As we study environmental biophysics, we will find that people from "primitive" cultures, and even animals, often have a far better understanding of the application of its principles than we do. Understanding the environment and how best to interact with it often makes the difference between life and death for them, whereas for us it may just mean a minor annoyance or an increased fuel bill.

1.8 Units

Units consistent with the Systeme International (SI) will be used in this book. The SI base units and their accepted symbols are the meter (m) for length, the kilogram (kg) for mass, the second (s) for time, the Kelvin (K) for thermodynamic temperature, and the mole (mol) for the amount of substance. Units derived from these, which we use in this book are given in Table 1.1. Additional derived units can be found in Page and Vigoureux (1974).

The Celsius temperature scale is more convenient for some biophysical problems than the thermodynamic (Kelvin) scale. We will use both. By definition $C = K - 273.15$. Since the Celsius degree is the same size as the Kelvin degree, derived units with temperature in the denominator can be written as either C^{-1} or K^{-1}. For example, units for specific heat are either $J kg^{-1} C^{-1}$ or $J kg^{-1} K^{-1}$. To distinguish between the two temperature scales, we will use T in standard font for Celsius temperature, and in bold font (**T**) for Kelvin temperature. Some useful factors for converting to SI units can be found in Table A.4 in the Appendix.

TABLE 1.1. Examples of derived SI units and their symbols.

Quantity	Name	Symbol	SI base units	Derived Units
area	—	—	m^2	—
volume	—	—	m^3	—
velocity	—	—	$m\ s^{-1}$	—
density	—	—	$kg\ m^{-3}$	—
force	Newton	N	$m\ kg\ s^{-2}$	—
pressure-force/area	Pascal	Pa	$kg\ m^{-1}\ s^{-2}$	$N\ m^{-2}$
energy	joule	J	$m^2\ kg\ s^{-2}$	$N\ m$
chemical potential	—	—	$m^2\ s^{-2}$	$J\ kg^{-1}$
power	watt	W	$m^2\ kg\ s^{-3}$	$J\ s^{-1}$
concentration	—	—	mol	—
mol flux density	—	—	$mol\ m^{-2}\ s^{-1}$	—
heat flux density	—	—	$kg\ s^{-3}$	$W\ m^{-2}$
specific heat	—	—	$m^2\ s^{-2}\ K^{-1}$	$J\ kg^{-1}\ K^{-1}$

To make the numbers used with these units convenient, prefixes are attached indicating decimal multiples of the units. Accepted prefixes, symbols, and multiples are shown in Table 1.2. The use of prefix steps smaller than 10^3 is discouraged. We will make an exception in the use of the cm, since mm is too small to conveniently describe the sizes of things like leaves, and m is too large. Prefixes can be used with base units or derived units, but may not be used on units in the denominator of a derived unit (e.g., g/m^3 or mg/m^3 but not mg/cm^3). The one exception to this rule that we make is the use of kg, which may occur in the denominator because it is the fundamental mass unit. Note in Table 1.2 that powers of ten are often used to write very large or very small numbers. For example, the number 0.0074 can be written as 7.4×10^{-3} or 86400 can be written as 8.64×10^4.

Most of the numbers we use have associated units. Before doing any computations with these numbers, it is important to convert the units to base SI units, and to convert the numbers using the appropriate multiplier from Table 1.2. It is also extremely important to write the units with the associated numbers. The units can be manipulated just as the numbers are, using the rules of multiplication and division. The quantities, as well as the units, on two sides of an equation must balance. One of the most useful checks on the accuracy of an equation in physics or engineering is the check to see that units balance. A couple of examples may help to make this clear.

Example 1.1. The energy content of a popular breakfast cereal is 3.9 kcal/g. Convert this value to SI units (J/kg).

TABLE 1.2.   Accepted SI prefixed and symbols for multiples and submultiples of units.

Multiplication Factor	Prefix	Symbol
1 000 000 000 000 000 000=10^{18}	exa	E
1 000 000 000 000 000=10^{15}	peta	P
1 000 000 000 000=10^{12}	tera	T
1 000 000 000=10^{9}	giga	G
1 000 000=10^{6}	mega	M
1 000=10^{3}	kilo	k
100=10^{2}	hecto	h
10=10^{1}	deka	da
0.1=10^{-1}	deci	d
0.01=10^{-2}	centi	c
0.001=10^{-3}	milli	m
0.000 001=10^{-6}	micro	μ
0.000 000 001=10^{-9}	nano	n
0.000 000 000 001=10^{-12}	pico	p
0.000 000 000 000 001=10^{-15}	femto	f
0.000 000 000 000 000 001=10^{-18}	atto	a

Solution.   Table A.4 gives the conversion, 1 J = 0.2388 cal so

$$\frac{3.9 \text{ kcal}}{\text{g}} \times \frac{10^3 \text{ cal}}{\text{kcal}} \times \frac{10^3 \text{ g}}{\text{kg}} \times \frac{1 \text{ J}}{0.2388 \text{ cal}} = 16.3 \times 10^6 \frac{\text{J}}{\text{kg}}$$

$$= 16.3 \text{ MJ/kg}.$$

Example 1.2.   Chapter 2 gives a formula for computing the damping depth of temperature fluctuations in soil as $D = \sqrt{\frac{2\kappa}{\omega}}$, where κ is the thermal diffusivity of the soil and ω is the angular frequency of temperature fluctuations at the surface. Figure 8.4 shows that a typical diffusivity for soil is around $0.4 \text{ mm}^2/\text{s}$. Find the diurnal damping depth.

Solution.   The angular frequency is $2\pi/P$, where P is the period of temperature fluctuations. For diurnal variations, the period is one day (see Chs. 2 & 8 for more details) so $\omega = 2\pi/1$ day. Converting ω and κ to SI base units gives:

$$\omega = \frac{2\pi}{1 \text{ day}} \times \frac{1 \text{ day}}{24 \text{ hr}} \times \frac{1 \text{ hr}}{60 \text{ min}} \times \frac{1 \text{ min}}{60 \text{ s}} = 7.3 \times 10^{-5} \text{ s}^{-1}$$

$$\kappa = \frac{0.4 \text{ mm}^2}{\text{s}} \times \frac{1 \text{ m}}{1000 \text{ mm}} \times \frac{1 \text{ m}}{1000 \text{ mm}} = 4 \times 10^{-7} \text{ m}^2/\text{s}$$

$$D = \sqrt{\frac{2 \times 4 \times 10^{-7} \text{ m}^2\text{s}^{-1}}{7.3 \times 10^{-5} \text{ s}^{-1}}} = 0.1 \text{ m}$$

Example 1.3. Units for water potential are J/kg (see Ch. 4). The gravitational component of water potential is calculated from $\psi_g = -gz$ where g is the gravitational constant (9.8 ms^{-2}) and z is height (m) above a reference plane. Reconcile the units on the two sides of the equation.

Solution. Note from Table 1.1 that base units for the joule are $\text{kg m}^2 \text{ s}^{-2}$ so

$$\frac{J}{kg} = \frac{kg \, m^2 \, s^{-2}}{kg} = \frac{m^2}{s^2}.$$

The units for the product, gz are therefore the same as the units for ψ.

Confusion with units is minimized if the numbers which appear within mathematical operators ($\sqrt{\ }$, exp, ln, sin, cos, tan, etc.) are dimensionless. In most cases we eliminate units within operators, but with some empirical equations it is most convenient to retain units within the operator. In these cases, particular care must be given to specifying the units of the equation parameters and the result. For example, in Ch. 7 we compute the thermal boundary layer resistance of a flat surface from

$$r_{Ha} = 7.4\sqrt{\frac{d}{u}} \tag{1.3}$$

where d is the length of the surface in m, u is the wind speed across the surface in m/s, and r_{Ha} is the boundary layer resistance in $\text{m}^2 \text{ s/mol}$. The constant 7.4 is the numerical result of evaluating numerous coefficients that can reasonably be represented by constant values. The constant has units of $\text{m}^2 \text{ s}^{1/2}/\text{mol}$, but this is not readily apparent from the equation. If one were to rigorously cancel units in Eq. (1.3) without realizing that the 7.4 constant has units, the result would appear to be an incorrect set of units for resistance. It would be a more serious matter if d were entered, for example, in mm, or u in cm/s, since then the result would be wrong. Whenever empirical equations like Eq. (1.3) are used in this book, we assume that parameters (u and d in the equation) are in SI base units, and we will specify the units of the result. This should avoid any ambiguity.

One other source of confusion can arise when units appear to cancel, leaving a number apparently dimensionless, but the units remain important to interpretation and use of the number. For example, the water content of a material might be reported as 0.29, or 29%. However, a water content of $0.29 \text{ m}^3/\text{m}^3$ can be quite different from a water content of 0.29 kg/kg. This type of confusion can always be eliminated by stating the units, even when they appear to cancel. In this book we use mole fraction, or mol/mol to express gas concentration. These units, though appearing to cancel, really represent moles of the particular gas, say water vapor, per mole of air. We therefore retain the mol/mol units with the numbers. It is often helpful to write out mol H_2O or mol air so that one is not tempted to cancel units which should not be canceled. This notation, however, tends to become cumbersome, and therefore is generally not used in the book.

References

Page, C.H. and P. Vigoureux (1974) The International System of Units, Nat. Bur. Stand. Spec. Publ. 330 U.S. Govt. Printing Office, Washington, D.C.

Seeger, R.J. (1973) Benjamin Franklin, New World Physicist, New York: Pergamon Press.

Problems

1.1. Explain why a concrete floor feels colder to you than a carpeted floor, even though both are at the same temperature. Would a snake or cockroach (both poikilotherms) arrive at the same conclusion you do about which floor feels colder?

1.2. In what ways (there are four) is energy transferred between living organisms and their surroundings? Give a description of the physical process responsible for each, and an example of each.

1.3. Convert the following to SI base units: 300 km, 5 hours, 0.4 mm^2/s, 25 kPa, 30 cm/s, and 2 mm/min.

1.4. In the previous edition of this book, and in much of the older literature, boundary layer resistances were expressed in units of s/m. The units we will use are m^2 s/mol. To convert the old units to the new ones, divide them by the molar density of air (41.65 mol m^{-3} at 20° C and 101 kPa). If boundary layer resistance is reported to be 250 s/m, what is it in m^2 s/mol? What is the value of the constant in Eq. (1.3) if the result is to be in s/m?

Temperature

2

Rates of biochemical reactions within an organism are strongly dependent on its temperature. The rates of reactions may be doubled or tripled for each $10°$ C increase in temperature. Temperatures above or below critical values may result in denaturation of enzymes and death of the organism.

A living organism is seldom at thermal equilibrium with its microenvironment, so the environmental temperature is only one of the factors determining organism temperature. Other influences are fluxes of radiant and latent heat to and from the organism, heat storage, and resistance to sensible heat transfer between the organism and its surroundings. Even though environmental temperature is not the only factor determining organism temperature, it is nevertheless one of the most important. In this chapter we describe environmental temperature variation in the biosphere and discuss reasons for its observed characteristics. We also discuss methods for extrapolating and interpolating measured temperatures.

2.1 Typical Behavior of Atmospheric and Soil Temperature

If daily maximum and minimum temperatures were measured at various heights above and below the ground and then temperature were plotted on the horizontal axis with height on the vertical axis, graphs similar to Fig. 2.1 would be obtained. Radiant energy input and loss is at the soil or vegetation surface. As the surface gets warmer, heat is transferred away from the surface by convection to the air layers above and by conduction to the soil beneath the surface. Note that the temperature extremes occur at the surface, where temperatures may be 5 to $10°$ C different from temperatures at 1.5 m, the height of a standard meteorological observation. This emphasizes again that the microenvironment may differ substantially from the macroenvironment.

A typical air temperature versus time curve for a clear day is shown in Fig. 2.2. Temperatures measured a few centimeters below the soil surface would show a similar diurnal pattern. The maximum rate of solar heat input to the ground is around 12 hours.

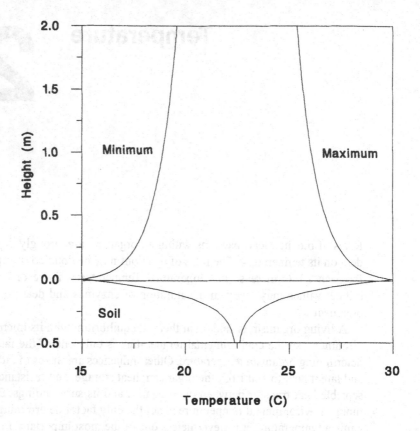

FIGURE 2.1. Hypothetical profiles of maximum and minimum temperature above and below soil surface on a clear, calm day.

The fact that the temperature maximum occurs after the time of maximum solar energy input is significant. This type of lag is typical of any system with storage and resistance to flow. Temperatures measured close to the exchange surface have less time lag and a larger amplitude than those farther from the surface. The principles involved can be illustrated by considering heat losses to a cold tile floor when you place your bare foot on it. The floor feels coldest (maximum heat flux to the floor) when your foot just comes in contact with it, but the floor reaches maximum temperature at a later time when heat flux is much lower.

The amplitude of the diurnal temperature wave becomes smaller with increasing distance from the exchange surface. For the soil, this is because heat is stored in each succeeding layer so less heat is passed on to the next layer. At depths greater than 50 cm or so, the diurnal temperature fluctuation in the soil is hardly measurable (Fig. 2.1).

The diurnal temperature wave penetrates much farther in the atmosphere than in the soil because heat transfer in the atmosphere is by eddy

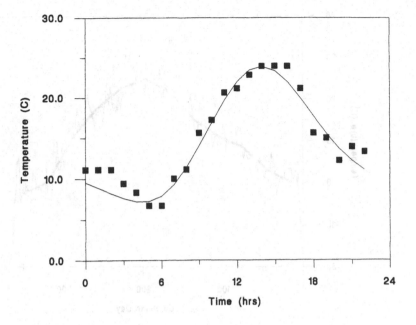

FIGURE 2.2. Hourly air temperature (points) on a clear fall day at Hanford, WA. The curve is used to interpolate daily maximum and minimum temperatures to obtain hourly estimates.

motion, or transport of parcels of hot or cold air over relatively long vertical distances, rather than by molecular motion. Within the first few meters of the atmosphere of the earth, the vertical distance over which eddies can transport heat is directly proportional to their height above the soil surface. The larger the transport distance, the more effective eddies are in transporting heat, so the air becomes increasingly well mixed as one moves away from the surface of the earth. This mixing evens out the temperature differences between layers. This is the reason for the shape of the air temperature profiles in Fig. 2.1. They are steep close to the surface because heat is transported only short distances by the small eddies. Farther from the surface the eddies are larger, so the change of temperature with height (temperature gradient) becomes much smaller.

In addition to the diurnal temperature cycle shown in Fig. 2.2, there also exists an annual cycle with a characteristic shape. The annual cycle of mean temperature shown in Fig. 2.3 is typical of high latitudes which have a distinct seasonal pattern from variation in solar radiation over the year. Note that the difference between maximum and minimum in Fig. 2.3 is similar to the difference between maximum and minimum of the diurnal cycle in Fig. 2.2. Also note that the time of maximum temperature (around day 200) significantly lags the time of maximum solar input (June 21; day 172). The explanation for this lag is the same as for the diurnal cycle.

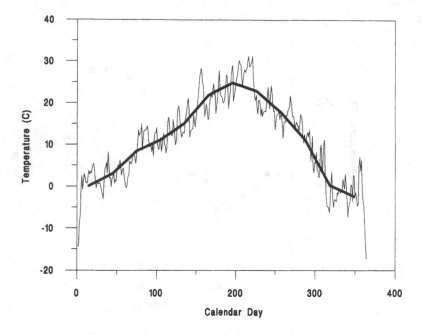

FIGURE 2.3. Daily average temperature variation at Hanford, WA for 1978. The heavy line shows the monthly mean temperatures.

2.2 Random Temperature Variation

In addition to the more or less predictable diurnal and annual temperature variations shown in Figs. 2.2 and 2.3, and the strong, predictable spatial variation in the vertical seen in Fig. 2.1, there are random variations, the details of which cannot be predicted. We can describe them using statistical measures (mean, variance, correlation etc.), but can not interpolate or extrapolate as we can with the annual, diurnal, and vertical variations. Figure 2.3 shows an example of these random variations. The long-term monthly mean temperature shows a consistent pattern, but the daily average temperature varies around this monthly mean in an unpredictable way. Figure 2.4 shows air temperature variation over an even shorter time. It covers a period of about a minute. Temperature was measured with a 25 μm diameter thermocouple thermometer.

The physical phenomena associated with the random variations seen in Figs. 2.3 and 2.4 make interesting subjects for study. For example, the daily variations seen in Fig. 2.3 are closely linked to weather patterns, cloud cover, and input of solar energy. The fluctuations in Fig. 2.4 are particularly interesting because they reflect the mechanism for heat transport in the lower atmosphere, and are responsible for some interesting optical phenomena in the atmosphere.

Since heat transfer in air is mainly by convection, or transport of parcels of hot or cold air, we might expect the air temperature at any instant to

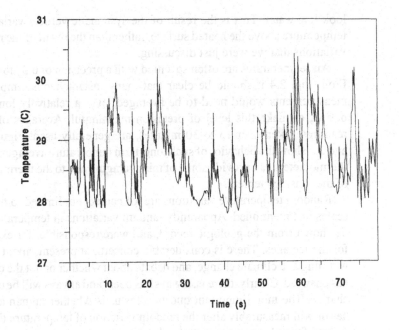

FIGURE 2.4. Air temperature 2 m above a desert surface at White Sands Missile Range, NM. Measurements were made near midday using a 25 μm diameter thermocouple.

differ substantially from the mean air temperature that one might measure with a large thermometer. The relatively smooth baseline in Fig. 2.4, with jagged interruptions, indicates a suspension of hot ascending parcels in a matrix of cooler, descending air. Well mixed air is subsiding, being heated at the soil surface, and breaking away from the surface as convective bubbles when local heating is sufficient.

Warm air is less dense than cold air, and therefore has a lower index of refraction. As light shines though the atmosphere, the hot and cold parcels of air act as natural lenses, causing the light to constructively and destructively interfere, giving rise to a diffraction pattern. Twinkling of stars and the scintillation of terrestrial light sources at night are the result of this phenomenon. The diffraction pattern is swept along with the wind, so you can look at the lights of a city on a clear night from some distance and estimate the wind speed and direction from the drift of the scintillation pattern.

So-called "heat waves" often seen on clear days also result from re-fractive index fluctuations (Lawrence et al., 1970). The drift of heat waves can be seen, and wind direction and speed can sometimes be estimated from the drift velocity. This phenomenon has been used to measure wind speed (Lawrence et al., 1972). More extreme heating at the surface can result in a mirage, where the heated, low-density air near the surface of the earth refracts the light from the sky to the observers eye, making land look like water. This is the result of the systematic vertical variation in temperature above the heated surface, rather than the result of the random variations that we were just discussing.

Air temperatures are often specified with a precision of 0.5° to 0.1° C. From Fig. 2.4 it should be clear that many instantaneous temperature measurements would need to be averaged, over a relatively long time period, to make this level of precision meaningful. Averages of many readings, taken over 15 to 30 minutes, are generally used. Figures 2.1 and 2.2 show the behavior of such long-term temperature averages. Large thermometers can provide some of this averaging due to the thermal mass of the sensing element.

Random temperature variations are, of course, not limited to the time scales just mentioned. Apparently random variations in temperature can be shown from the geologic record, and were responsible, for example, for the ice ages. There is considerable concern, at present, about global warming and climate change, and debate about whether or not the climate has changed. Clearly, there is, always has been, and always will be climate change. The more important question for us is whether human activity has or will measurably alter the random variation of temperature that has existed for as long as the earth has been here.

2.3 Modeling Vertical Variation in Air Temperature

The theory of turbulent transport, which we study in Ch. 7, specifies the shape of the temperature profile over a uniform surface with steady-state conditions. The temperature profile equation is:

$$T(z) = T_0 - \frac{H}{0.4 \hat{\rho} c_p u^*} \ln \frac{z-d}{z_H} \tag{2.1}$$

where $T(z)$ is the mean air temperature at height z, T_0 is the apparent aerodynamic surface temperature, z_H is a roughness parameter for heat transfer, H is the sensible heat flux from the surface to the air, $\hat{\rho} c_p$ is the volumetric specific heat of air (1200 J m^{-3} C^{-1} at 20° C and sea level), 0.4 is von Karman's constant, and u^* is the friction velocity (related to the friction or drag of the stationary surface on the moving air). The reference level from which z is measured is always somewhat arbitrary, and the correction factor d, called the zero-plane displacement, is used to adjust for this. For a flat, smooth surface, $d = 0$. For a uniformly vegetated surface, $z_H \simeq 0.02h$, and $d \simeq 0.6h$, where h is canopy height.

We derive Eq. (2.1) in Ch. 7, but use it here to interpret the shape of the temperature profile and extrapolate temperatures measured at one height to other heights. The following points can be made.

1. The temperature profile is logarithmic (a plot of $\ln(z-d)/z_H$ vs. $T(z)$ is a straight line).

2. Temperature increases with height when H is negative (heat flux to-ward the surface) and decreases with height when H is positive. During the day, sensible heat flux is generally away from the surface so T decreases with height.
3. The temperature gradient at a particular height increases in magnitude as the magnitude of H increases, and decreases as wind or turbulence increases.

Example 2.1. The following temperatures were measured above a 10 cm high alfalfa crop on a clear day. Find the aerodynamic surface temperature, T_0.

Height (m)	0.2	0.4	0.8	1.6
Temperature (C)	26	24	23	21

Solution. It can be seen from Eq. (2.1) that $T(z) = T_0$ when the ln term is zero, which happens when $z = d + z_H$ since $\ln(z_H/z_H) = \ln(1) = 0$. If $\ln[(z - d)/z_H]$ is plotted versus T (normally the independent variable is plotted on the abscissa or horizontal axis, but when the independent variable is height, it is plotted on the ordinate or vertical axis) and ex-trapolate to zero, the intercept will be T_0. For a 10 cm (0.1 m) high canopy, $z_H = 0.002$ m, and $d = 0.06$ m. The following can therefore be computed:

Height (m)	0.2	0.4	0.8	1.6
Temp. (C)	26	24	23	21
$(z - d)/z_H$	70	170	370	770
$\ln[(z - d)/z_H]$	4.25	5.14	5.91	6.65.

The Figure for Example 2.1 on the following page is plotted using this data, and also shows a straight line fitted by linear least squares through the data points that is extrapolated to zero on the log-height scale. The intercept is at 34.6° C, which is the aerodynamic surface temperature.

Example 2.2. The mean temperature at 5:00 hrs, 2 m above the soil surface is 3° C. At a height of 1 m, the temperature is 1° C. If the crop below the point where these temperatures are measured is 50 cm tall, will the crop experience a temperature below freezing?

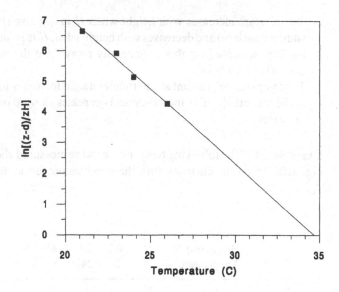

FIGURE FOR EXAMPLE 2.1.

Solution. This problem could be solved by plotting, as we did in Example 1, or it could be done algebraically. Here we use algebra. The constants, $H/(0.4\rho c_p u^*)$ in Eq. (2.1) are the same for all heights. For convenience, we represent them by the symbol A. Equation (2.1) can then be written for each height as

$$3 = T_0 - A \ln \frac{2 - 0.6 \times 0.5}{0.02 \times 0.5} = T_0 - 5.14A$$

$$1 = T_0 - A \ln \frac{1 - 0.6 \times 0.5}{0.02 \times 0.5} = T_0 - 4.25A.$$

Subtracting the second equation from the first, and solving for A gives $A = -2.25°$ C. Substituting this back into either equation gives $T_0 = -8.6°$ C. Knowing these, now solve for $T(h)$ where $h = 0.5$ m:

$$T(h) = -8.6 + 2.25 \ln \frac{0.5 - 0.6 \times 0.5}{0.5 \times 0.02} = -1.9 \, \text{C}.$$

So the top of the canopy is below the freezing temperature.

These two examples illustrate how temperatures can be interpolated or extrapolated. In each case, two temperatures are required, in addition to information about the height of roughness elements at the surface. Typically temperature is measured at a single height. From Eq. (2.1), it is clear that additional information about the sensible heat flux density H and the wind would be needed to extrapolate a single temperature measurement. This is taken up later when we have the additional tools needed to model heat flux.

2.4 Modeling Temporal Variation in Air Temperature

Historical weather data typically consist of measurements of daily maximum and minimum temperature measured at a height of approximately 1.5 m. There are a number of applications which require estimates of hourly temperature throughout a day. Of course, there is no way of knowing what the hourly temperatures were, but a best guess can be made by assuming that minimum temperatures normally occur just before sunrise and maximum temperatures normally occur about two hours after solar noon. The smooth curve in Fig. 2.2 shows this pattern. The smooth curve was derived by fitting two terms of a Fourier series to the average of many days of hourly temperature data which had been normalized so that the minimum was 0 and the maximum was 1. The dimensionless diurnal temperature function which we obtained in this way is:

$$\Gamma(t) = 0.44 - 0.46 \sin(\omega t + 0.9) + 0.11 \sin(2\omega t + 0.9) \qquad (2.2)$$

where $\omega = \pi/12$, and t is the time of day in hours ($t = 12$ at solar noon). Using this function, the temperature for any time of the day is given by

$$T(t) = T_{x,i-1}\Gamma(t) + T_{n,i}[1 - \Gamma(t)] \qquad 0 < t \le 5$$
$$T(t) = T_{x,i}\Gamma(t) + T_{n,i}[1 - \Gamma(t)] \qquad 5 < t \le 14 \qquad (2.3)$$
$$T(t) = T_{x,i}\Gamma(t) + T_{n,i+1}[1 - \Gamma(t)] \qquad 14 < t < 24.$$

Here, T_x is the daily maximum temperature and T_n is the minimum temperature. The subscript i represents the present day; $i - 1$ is the previous day, and $i + 1$ is the next day. The curve in Fig. 2.2 was drawn using Eqs. (2.2) and (2.3). Note that t is solar time. The local clock time usually differs from solar time, so adjustments must be made. These are discussed in detail in a later chapter.

Example 2.3. Estimate the temperature at 10 AM on a day when the minimum was 5° C and the maximum was 23° C.

Solution. At $t = 10$ hrs, (note that angles are in radians)

$$\Gamma = 0.44 - 0.46 \sin\left(\frac{3.14 \times 10}{12} + 0.9\right)$$

$$+ 0.11 \sin\left(\frac{2 \times 3.14 \times 10}{12} + 0.9\right) = 0.593.$$

Since 10 is between 5 and 14, the middle form of Eq. (2.3) is used, so

$$T(10) = 23 \times 0.593 + 5 \times (1 - 0.593) = 15.7.$$

2.5 Soil Temperature Changes with Depth and Time

The temperature of the soil environment is also important to many living organisms. Figure 2.1 shows a typical range for diurnal temperature variation with depth in soil. The features to note are that the temperature extremes occur at the surface where radiant energy exchange occurs, and that the diurnal variation decreases rapidly with depth in the soil. Figure 2.5 also shows these relationships and gives additional insight into soil temperature variations. Here, temperatures measured at three depths are shown.

Note that the diurnal variation is approximately sinusoidal, that the amplitude decreases rapidly with depth in the soil, and that the time of maximum and minimum shifts with depth. At the surface, the time of maximum temperature is around 14:00 hours, as it is in the air. At deeper depths the times of the maxima and minima lag solar noon even farther, and at 30 to 40 cm, the maximum is about 12 hours after the maximum at the surface.

We derive equations for heat flow and soil temperature later when we discuss conductive heat transfer. For the moment, we just give the equation which models temperatures in the soil when the temperature at the surface is known. This model assumes uniform soil properties throughout the soil profile and a sinusoidally varying surface temperature. Given these assumptions, the following equation gives the temperature as a function of depth and time:

$$T(z, t) = T_{ave} + A(0) \exp(-z/D) \sin[\omega(t - 8) - z/D] \qquad (2.4)$$

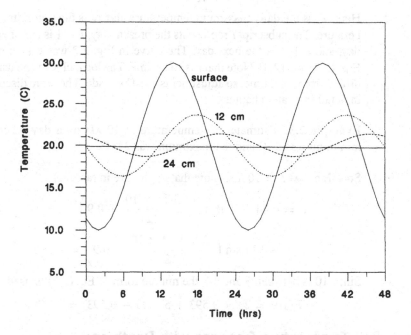

FIGURE 2.5. Hypothetical temperature variations in a uniform soil at the surface and two depths showing the attenuation of diurnal variations and the shift in maxima and minima with depth.

where T_{ave} is the mean daily soil surface temperature, ω is $\pi/12$, as in Eq. (2.2), $A(0)$ is the amplitude of the temperature fluctuations at the surface (half of the peak-to-peak variation) and D is called the damping depth. The "-8" in the sine function is a phase adjustment to the time variable so that when time $t = 8$, the sine of the quantity in brackets is zero at the surface ($z = 0$). We discuss computation of diurnal damping depth in Ch. 8. It has a value around 0.1 m for moist, mineral soils, and 0.03 to 0.06 m for dry mineral soils and organic soils.

In many cases we are not interested in the time dependence of the soil temperature, but would just like to know the range of temperatures at a particular depth. It is known that the range of the sine function is -1 to 1 so Eq. (2.4) gives the range of soil temperature variation as

$$T(z) = T_{ave} \pm A(0) \exp(-z/D) \qquad (2.5)$$

where the $+$ gives the maximum temperatures and the $-$ the minimum.

Example 2.4. At what depth is the soil temperature within $\pm 0.5°$ C of the mean daily surface temperature if the temperature variation at the surface (amplitude) is $\pm 15°$ C?

Solution. The amplitude of the desired temperature variation is $0.5°$ C. Rearranging Eq. (2.5) and taking the logarithm of both sides gives

$$\frac{z}{D} = -\ln \frac{T(z) - T_{ave}}{A(0)} = -\ln \frac{0.5}{15} = 3.4.$$

If $D = 12$ cm, then the depth for diurnal variations less than $\pm 0.5°$ C would be 3.4×12 cm $= 41$ cm. Therefore a depth of at least 40 cm needs to be dug to obtain a soil temperature measurement that is not influenced by the time of day the temperature is measured.

The annual soil temperature pattern is similar to the diurnal one, but with a much lower frequency and a much larger damping depth. Equations (2.4) and (2.5) are used to describe the annual variation, but D is around 2 m, and ω is $2\pi/365$ days.

While Eqs. (2.4) and (2.5) are useful relationships for getting a general idea of how soil temperature varies with depth and time, it is important to remember their limitations. The thermal properties do vary with depth, and the temperature variation at the surface is not necessarily sinusoidal. Temperature variations over periods longer than a day or a year also have an effect. In spite of these limitations, however, a lot can be learned from this simple model.

Clearly, from Eq. (2.4), the value of the damping depth D is key to predicting the penetration into the soil of a temperature variation at the surface. Data such as that in Fig. 2.5 can be used to estimate D. Solving Eq. (2.5) for $T(z) - T_{ave}$ and applying it at two depths permits solution for D. If the amplitude of the temperature wave is $T(z_1) - T_{ave} = A_1$ at one depth and is $T(z_2) - T_{ave} = A_2$ at a second depth, then

$$D = \frac{z_1 - z_2}{\ln(A_2) - \ln(A_1)}.$$

2.6 Temperature and Biological Development

Now that we have some idea of the behavior of temperature in the natural environment of living organisms, we want to consider temperature from a biological perspective. For this analysis, we assume that temperatures of plants, microbes, and insects are the same as the temperature of their environment. We need to remember, however, that such is generally not the case. Later we develop the tools to compute organism temperature from environmental temperature and can then consider what effect this will have on the organism.

Temperature strongly influences the rates of all metabolic processes in living organisms, and therefore affects almost all aspects of the growth and development of an organism. Here we want to consider the effect of temperature on the rate of development. We define development as the orderly progress of an organism through defined stages from germination to death. Development differs from growth, which we define as the accumulation of dry matter. Developmental stages vary, depending on the organism being described. In plants, stages such as germination, emergence, leaf appearance, flowering, and maturity can be defined, as can intermediate stages within many of these stages. In insects, stages such as egg, larva, and adult can be identified, and with other living organisms developmental stages can be similarly identified and defined.

Figure 2.6 shows the time taken for completion of the egg stage of Dacus cucurbitae at constant temperatures ranging from 10° to 35° C. Development time is short at temperatures between 20° and 30° C, but increases markedly at both higher and lower temperatures. Above 37° C and below 15° C, development times are very long. We are interested in determining the time taken for completion of the egg stage (or some other developmental stage) under varying temperature conditions. This can be found by computing the reciprocals of the times in Fig. 2.6 to obtain a rate of development. Figure 2.7 shows the rate of development (with units of completed stages per day) as a function of temperature. The shape of this curve is similar for many biological processes, and has been described mathematically using the theory of rate processes (Sharpe and DeMichele, 1977; Wagner, et al. 1984). Such detailed models are written in terms of three exponentials, and are therefore difficult to both fit and compute. It is evident, however, that the data in Fig. 2.7 are closely approximated by two straight lines. Again, this is typical of many biological responses to temperature.

Descriptions of the rate of development, such as Fig. 2.7, are the basis for determining the time taken to complete a developmental process when temperature varies. Assume, for example, that one has measurements of soil temperature, and wishes to predict the time required to complete the

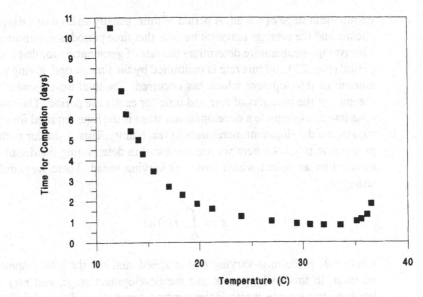

FIGURE 2.6. Time for development of melon fly (*Dacus cucurbitae*) eggs at different temperatures.

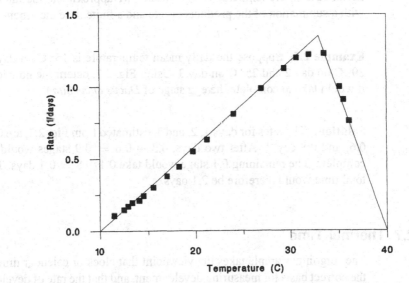

FIGURE 2.7. Development rate of melon fly eggs showing the almost linear response to temperature.

germination stage of a seed. A period of time, usually an hour or a day, is chosen and the average temperature over that time period is determined. The average temperature determines the rate of germination for that time period (Fig. 2.7) and this rate is multiplied by the time period, giving the amount of development which has occurred. The total development is the sum of the products of rate and time for each time period. The total time taken to complete a developmental stage is the time required for the sum of the development increments to reach unity. This is similar to the problem in physics where we are interested in determining the distance traveled by an object which moves at varying speed. There we would write

$$s = \int_{t_1}^{t_2} r(t)\, dt$$

where $r(t)$ is the time-varying rate or speed, and s is the total distance traveled. In this analogy, s is like the development stage, and $r(t)$ is the development rate, which is temperature-dependent and may therefore vary in some arbitrary way with time. Since the functional form for the rate is generally not known for development calculations (except in the trivial case where temperature is constant) we approximate the integral with a summation of the products of rate and a finite time increment.

Example 2.5. Suppose the daily mean temperature is 15° C on day 1, 20° C on day 2 and 25° C on day 3. Using Fig. 2.7, determine how long it would take to complete the egg stage of *Dacus cucurbitae*.

Solution. The rates for days 1, 2, and 3, estimated from Fig. 2.7, are 0.3, 0.6, and 0.8 day^{-1}. After two days, $0.3 + 0.6 = 0.9$ stages would be complete. The remaining 0.1 stage would take $0.1/0.8 \simeq 0.1$ days. The total time would therefore be 2.1 days.

2.7 Thermal Time

The forgoing example takes the viewpoint that clock or calendar time is the correct basis for measuring development, and that the rate of development of an ectotherm (an organism whose temperature is environmentally determined) varies depending on environmental temperature. Another viewpoint is that there exists a time scale in which the rate of development of organisms is constant, and information like that in Fig. 2.7 provides a means of transforming biological time to clock or calendar time. Monteith (1977) uses the term *thermal time* to describe a time scale in which the development rate of organisms is constant. It has also been referred to as physiological time or p-time. Units of thermal time are day-degrees or hour-degrees. Units for p-time are p-days or p-hours.

The formal transforms which convert one time scale to the other, for an organism whose development rate depends only on temperature, is

$$\tau = \int_0^t R[T(t)]\, dt$$

$$t = \int_0^\tau g(T)\, d\tau \tag{2.6}$$

where τ is the thermal time and R is the rate of development at temperature T (which, in turn, depends on time). The function g is the inverse of R and allows, in principle, the conversion of thermal time back to clock time.

In practice, the integral in Eq. (2.6) is always approximated as a sum because temperature generally is not a predictable function of time. For the usual calculation of thermal time we assume a straight line relationship between development rate and temperature, such as that shown in Fig. 2.7. We also assume that temperatures are always within the range $T_b \ldots T_m$, where T_b is the base temperature (low temperature at which development stops) and T_m is the temperature at which the development rate is maximum. Thermal time, and therefore organism development, is then directly proportional to the sum of products of $(T_i - T_b)$ and the length of the time increment, where T_i is the temperature at a particular time, with the condition that $T_i - T_b > 0$. Given these assumptions, the equation for thermal time increments is

$$\Delta \tau_i = (T_i - T_b)\Delta t \quad \text{when} \quad T_i > T_b; \quad \text{otherwise} \quad \Delta \tau_i = 0. \tag{2.7}$$

The time step, Δt, is chosen so that temperature is fairly constant during one time increment. The units of $\Delta \tau$ are day-degrees, or hour-degrees, depending on the units of Δt. No thermal time is accumulated when T_i is at or below the base temperature. Thermal time is computed as:

$$\tau_n = \sum_{i=1}^n \Delta \tau_i. \tag{2.8}$$

From Fig. 2.7, it can be seen that the rate is $1.35\ \mathrm{d^{-1}}$ when the temperature is $T_m = 33^\circ$ C. The base temperature is $T_b = 10^\circ$ C. At 33° C, the time for completion is $1/1.35 = 0.74$ days. The thermal time for completion at this constant temperature is 0.74 days, or $(33^\circ\,\mathrm{C} - 10^\circ\,\mathrm{C})/1.35 = 17.0$ day-degrees. When the temperature of the melon fly eggs varies during germination, we can use the varying temperature, with Eq. (2.8), to find τ_n since the start of the stage. Once τ_n reaches 17.0 day-degrees, the stage will be complete.

The inverse operation indicated by the second of Eqs. (2.6) is used to find the calendar or clock time required to complete a developmental stage. An analytical form of the inverse is not possible except in the trivial case where temperature is constant. To find the calendar time required for completion of the egg stage in the example just presented, we would construct a table of t_n and the corresponding τ_n. We would then enter the table at $\tau_n = 17.0$ day-degrees and find how many calendar days were required to reach that value.

The term heat unit has been used in connection with the day-degree, but this is clearly inappropriate. The unit has nothing to do with heat or its accumulation, but defines a quantity which bears a simple linear relationship to biological time.

2.8 Calculating Thermal Time from Weather Data

Reports of thermal time for predicting crop or pest development are generally based on calculations from daily maximum (T_x) and minimum (T_n) temperatures using

$$\tau_n = \sum_{i=1}^{n} \left(\frac{T_{xi} + T_{ni}}{2} - T_b \right) \Delta t. \tag{2.9}$$

If the average of the maximum and minimum temperatures is less than the base temperature or greater than some maximum temperature, zero is added to the sum for that day. Several assumptions are implicit in using Eq. (2.9):

1. the growing region of the plant is at air temperature
2. the hourly air temperature does not go below the base temperature or above the maximum temperature during a day
3. the process being predicted is linear with temperature between the base and maximum temperatures.

The time increment, Δt, is taken as one day. The progress toward completion of a developmental stage is reported in day-degrees above a specified base temperature. Day-degrees required for completion of a developmental stage are used to determine completion or progress toward completion of development. The role of extreme temperatures in calculation of day-degrees is discussed in the next section. Errors from the growing point temperature not being at air temperature can be significant. For example, the growing point in corn is below the soil surface in early developmental stages, and failure to use soil temperature during this time can result in errors of five days or more in predictions of tasselleling date.

The base temperature and thermal time requirements of organisms depend, of course, on species and developmental stage. There is some evidence, however, that base temperatures may be relatively constant for developmental processes within a species and genotype. Angus et al. (1981) report the base temperatures of 30 species, including both temperate and tropical crops. Selected values are shown in Table 2.1. Note that the base temperatures fall into two groups, one centered around 2° C, and the other around 11° C. The former are representative of temperate species such as wheat, barley, pea, etc., and the latter of tropical crops such as maize, millet, and sorghum. Base temperatures and thermal time requirements can be estimated using the values from Table 2.1, but it should be recognized that considerable genotypic variability ex-

TABLE 2.1. Base temperature and thermal time requirement for emergence of selected temperate and tropical crops (from Angus et al. 1980).

Species	T_b (C)	day-deg	Species	T_b (C)	day-deg
wheat	2.6	78	maize	9.8	61
barley	2.6	79	pearl millet	11.8	40
oats	2.2	91	sorghum	10.6	48
field pea	1.4	110	peanut	13.3	76
lentil	1.4	90	cowpea	11.0	43
rape	2.6	79	pigeon pea	12.8	58

ists. This variability is very useful in fitting specific genotypes to specific environments.

The thermal time concept has been useful in applications where one wishes to predict harvest dates or emergence dates from planting dates, for finding varieties (with known thermal time requirements) which are best suited to a given climate, and for predicting disease or insect development.

Example 2.6. Using the weather data in the first three columns of the following table, and assuming the seed temperature is the same as the air temperature, how long would it take to germinate seed of cowpea which was planted on day 188?

Solution. From Table 2.1, cowpea has a base temperature of 11° C, and a thermal time requirement of 43 day degrees for emergence. Column 4 in the table is the quantity in brackets in Eq. (2.9), and column 5 is the summation according to Eq. (2.9). Emergence takes place on day 199 since the thermal time requirement is completed in that day.

day	T_{max}	T_{min}	$T_{ave} - T_b$	τ_n
188	23.3	12.2	——	0
189	23.9	9.4	5.65	5.65
190	17.2	6.1	0.65	6.3
191	21.1	7.8	3.45	9.75
192	23.3	10.6	5.95	15.7
193	29.4	12.8	10.10	25.8
194	22.8	13.3	7.05	32.85
195	15.0	5.6	−0.7	32.85
196	18.9	6.7	1.8	34.65
197	17.2	10.0	2.6	37.25
198	20.0	8.3	3.15	40.4
199	25.6	10.0	6.8	47.2

2.9 Temperature Extremes and the Computation of Thermal Time

Equation (2.9) uses the daily mean temperature to compute the thermal time increment for a given day. The temperature during the diurnal cycle has varied, however, and may have been outside the linear portion of the temperature response function, even though the mean temperature for the day was within that range. The correct estimate of thermal time would be obtained by shortening Δt to one hour, and summing hour-degrees to determine thermal time for the day. This is often done with insect models, where development times are short and good precision is required. Even when only daily maximum and minimum temperatures are known, the hourly values can be estimated using the interpolation method discussed earlier.

Another problem arises when temperatures are high. The computations of thermal time which have just been considered apply only for temperatures below T_m, the temperature where the development rate is maximum. At temperatures above T_m, Eq. (2.9) predicts that the development rate will continue to increase, while Fig. 2.7 shows that, in fact, it decreases. Equation (2.6), however, is very general, and includes any possible function of temperature. Therefore, a high temperature cutoff could be included, as well as a base temperature, in Eq. (2.7) to obtain:

$$\Delta \tau_i = 0 \quad \text{when} \quad T_i \leq T_b$$
$$\Delta \tau_i = (T_i - T_b)\Delta t \quad \text{when} \quad T_b < T_i < T_m$$
$$\Delta \tau_i = \frac{T_x - T_i}{T_x - T_m}(T_m - T_b)\Delta t \quad \text{when} \quad T_m \leq T_i < T_x \tag{2.10}$$
$$\Delta \tau_i = 0 \quad \text{when} \quad T_x \leq T_i.$$

Here T_x is the maximum temperature at which development can occur, and the high temperature response, as well as the low, has been approximated by a linear function.

2.10 Normalization of Thermal Time

Our choice of the day-degree as a unit for measuring physiological time is completely arbitrary, and is used mainly for historical reasons. Empirical relationships between accumulated day-degrees and development were found long before the physiological basis for these relationships was discovered. Since a linear relationship exists between day-degrees and development, it is convenient to use the day-degree time scale to measure progress of organism development. For a given organism at a given stage of development, rate of development is constant when measured in day-degrees. However, any other measure of time which is linearly related to accumulated day-degrees would work as well, and might sometimes be even better.

One such time scale is obtained by dividing each of the rates in Fig. 2.7 by the maximum rate (rate at T_m). All of the resulting rates then become dimensionless and represent the rate of development relative to the rate under optimal conditions. If time is considered as advancing at the rate of 1 pday per 0.74 day (or phour per hour) under optimum conditions (calculated at 33° C from Fig. 2.7), then time advances under suboptimal conditions at a rate less than 1.35 pday/day (remember, pday is a physiological day). The summation of these, possibly suboptimal pdays/day (as determined by temperature), is a direct measure of the accumulation of calendar time toward completion of a process. If maturity of some organism requires 50 days under optimal conditions (50 pdays, therefore), then, if the temperature is such that the organism accumulates only 0.4 pdays/day (for example 18° C in Fig. 2.7), maturity will require $50/0.4 = 125$ calendar days. This normalization of thermal time so that the response is dimensionless and ranges from 0 to 1 allows the generalization of the thermal time concept to other environmental variables. This generalization can be an extremely powerful tool for modeling the response of organisms to their environment.

2.11 Thermal Time in Relation To Other Environmental Variables

The development rate concept can be extended to other environmental variables which alter the relationship between development and temperature. For example, the rate of completion of budburst in a number of northern temperate tree species depends on winter chilling (Cannell and Smith, 1983). Other examples are the vernalization requirement for reproductive growth of winter wheat (Porter, 1983; Weir et al., 1984) and photoperiod requirements for development in many species. Temperature and leaf wetness both affect the development of foliar diseases of plants, so a thermal time modified by a leaf wetness factor determines the rate of development.

To predict development when two or more environmental variables affect development rate, we need to determine rate curves for each combination of conditions. This is sometimes quite simple. For example, in the infection of a plant by organisms which can only grow when the leaf surface is wet, development rate is zero when leaves are dry, and progresses at the temperature-determined rate when the leaf surface is wet. For photoperiod and chill requirements, the calculations are somewhat more involved, since the development rate is dependent on the chill or photoperiod.

As an example, consider a plant which flowers under long-day conditions, but not when days are short. The photothermal time is computed from

$$\tau_n = \sum_{i=1}^{n} f(l_d)\Delta\tau_i \qquad (2.11)$$

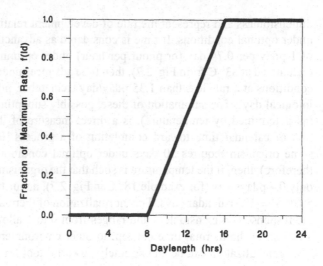

FIGURE 2.8. Photoperiod response function for a long-day species.

where l_d is the day length. A daylength response function for a long-day plant is shown in Fig. 2.8. When the daylength is shorter than eight hours, no development occurs. For days longer than 16 hours, development occurs at the temperature determined rate. For lengths between eight and 16 hrs, Fig. 2.8 gives the factor to multiply the thermal time increment by to determine the advance of photothermal time.

Chill moderated thermal time is computed similarly. As the plant experiences temperatures near freezing, chill units accumulate. A function similar to Fig. 2.8, but with chill units as the independent variable, determines the rate of accumulation of thermal-chill time.

The use of relative rates for thermal time and for the chill, moisture, and photoperiod factors which moderate thermal time, is attractive for several reasons. It provides a generally applicable approach to modeling effects of temperature on development and minimizes the number of variables needed to describe the temperature response. It also provides a simple bridge from field to laboratory time scales. From laboratory experiments, the minimum time required for a given developmental process to occur may be known. From field temperature (and other environmental data, when applicable) the relative rate curves can be used to determine the number of field days which are equivalent to one laboratory day at optimum conditions.

References

Angus, J. F., R. B. Cunningham, M. W. Moneur, and D. H. MacKenzie. (1981) Phasic development in field crops. I. Thermal response in the seedling stage. Field Crops Res. 3:365-378.

Cannell, M. G. R. and R. I. Smith. (1983) Thermal time, chill days and prediction of budburst in *Picea sitchensis*. J. Appl. Ecol. 20:951-963.

Lawrence, R.S., G.R. Ochs, and S.F. Clifford. (1970) Measurements of atmospheric turbulence relevant to optical propagation, J. Opt. Soc. Am. 60:826-830.

Lawrence, R.S., G.R. Ochs, and S.F. Clifford. (1972) Use of scintillations to measure average wind across a light beam. Appl. Optics 11:239–243.

Monteith, J. (1977) Climate and the efficiency of crop production in Britain. Phil. Trans. R. Soc. Lond. B281:277–294.

Porter, J. R. (1983) A model of canopy development in winter wheat. J. Agric. Sci., Camb. 102:283–292.

Sharpe, P. J. H. and D. W. DeMichele. (1977) Reaction kinetics of poikilotherm development. J. Theor. Biol. 64:649–670.

Wagner, T. L., H. Wu, P. J. H. Sharpe, R. M. Schoolfield, and R. N. Coulson. (1984) Modeling insect development rates: a literature review and application of a biophysical model. Ann. Entomol. Soc. Am. 77:208–225.

Weir, A. H., P. L. Bragg, J. R. Porter, and J. H. Rayner. (1984) A winter wheat crop simulation model without water or nutrient limitations. J. Agric. Sci., Camb. 102:371–382.

Problems

2.1. Using the midday temperature data in the following table:

a. Plot height as a function of mean temperature.
b. Plot $\ln[(z - d)/z_H]$ (Eq. (2.1)) as a function of mean temperature; assume the canopy height is $h = 0.15$ m.
c. From the plot in b, find the aerodynamic surface temperature, T_0.

Height (m)	Mean air temperature (C)
6.4	31.08
3.2	31.72
1.6	32.37
0.8	33.05
0.4	33.80
0.2	34.74
0.1	36.91

d. If $u^* = 0.2$ m s^{-1}, what is the sensible heat flux, H?

2.2. Find the damping depth, D, from the data in Fig. 2.5.

2.3. If the daily maximum and minimum soil temperatures at the soil surface are 35° C and 15° C, respectively for several consecutive days, plot the maximum and minimum temperatures as a function of depth to 30 cm for a soil with damping depth of 10 cm.

2.4. Using the weather data in the following table, predict the date of flowering of spring wheat planted on day 119 if flowering requires 900 day-degrees from planting with a base temperature of $3°$ C. How much later will flowering occur if planting is delayed to day 150?

day	Tx	Tn	day	Tx	Tn	day	Tx	Tn
119	5.	−0.6	153	18.9	6.1	187	19.4	6.1
120	11.7	1.7	154	17.8	5.6	188	22.2	10.6
121	21.1	4.4	155	15.	4.4	189	22.2	9.4
122	18.9	3.3	156	13.3	6.7	190	21.7	12.2
123	17.8	3.9	157	14.4	5.6	191	25.	10.
124	6.	−1.1	158	15.6	3.9	192	27.2	11.1
125	10.6	−1.7	159	17.2	1.7	193	28.9	16.1
126	14.4	1.7	160	20.	6.1	194	27.2	11.1
127	17.8	5.6	161	23.9	6.1	195	29.4	12.2
128	13.9	2.2	162	25.6	6.7	196	18.3	6.1
129	12.8	2.2	163	27.8	8.9	197	14.4	6.1
130	6.	−1.7	164	19.4	11.7	198	18.3	3.9
131	15.	−1.1	165	17.2	12.2	199	23.9	11.1
132	18.3	5.	166	22.8	10.6	200	28.3	13.3
133	17.8	2.8	167	27.8	11.1	201	28.3	12.8
134	21.1	9.4	168	28.3	11.7	202	30.	13.3
135	20.6	9.4	169	28.3	12.8	203	26.1	3.9
136	17.8	7.2	170	28.9	8.9	204	22.8	3.9
137	21.7	10.	171	30.6	11.1	205	25.6	5.
138	15.6	4.4	172	31.7	13.3	206	28.9	9.4
139	13.9	3.3	173	26.1	13.9	207	31.1	10.6
140	16.7	5.	174	27.8	11.7	208	32.2	9.4
141	21.7	5.	175	28.9	14.4	209	33.9	15.
142	23.9	10.	176	29.4	12.2	210	33.9	13.3
143	19.4	5.	177	22.2	15.	211	34.4	16.1
144	17.8	5.6	178	28.9	12.8	212	34.4	12.8
145	23.9	11.1	179	21.1	11.7	213	28.9	8.3
146	25.	6.7	180	14.4	12.2	214	22.8	8.9
147	15.	0.	181	19.4	11.7	215	22.2	11.1
148	12.2	2.8	182	23.3	13.3	216	21.7	11.1
149	12.2	4.4	183	16.7	8.3	217	24.4	7.8
150	16.1	3.3	184	21.1	9.4	218	27.8	8.9
151	18.3	6.7	185	18.3	7.8	219	31.7	11.7
152	22.2	5.6	186	18.3	8.3			

Water Vapor and Other Gases

3

Terrestrial organisms live in a gaseous medium composed mostly of nitrogen and oxygen. Water vapor is present in varying amounts, and carbon dioxide and other gases, in trace amounts. Organisms exchange oxygen, carbon dioxide, and water vapor with their surroundings. Carbon dioxide is a substrate for photosynthesis and oxygen is a product; while oxygen is a substrate for respiration and carbon dioxide is a product. Exchange of these gases with the environment is therefore a requirement for life. Water vapor almost always moves from the organism to the environment. The humidity of the organism is near 100 percent, while the surroundings are nearly always much drier. The organism must remain in a highly hydrated state in order for biochemical reactions to occur, so the constant loss of water is a threat to survival, and frequent access to liquid water is a necessity for most terrestrial organisms. The intake of liquid water and the loss of water vapor to the environment are usually the most important components of the water budget of an organism.

Water loss is generally viewed as detrimental to the organism, though it may have some benefit in the circulation system of plants. It does, however, have a definite benefit when we consider the energy balance of the organism. Environmental temperatures are often higher than can be tolerated by biological systems. If there were not some mechanism for cooling the organism, it would perish. As water evaporates, roughly 44 kilojoules of energy are required to convert each mole to the vapor state. This is called the latent heat of vaporization. It is 580 times the energy required to change the temperature of one mole of water by one Celsius degree, and therefore represents an enormous sink for energy in the organism environment. Evaporative cooling is a natural way of controlling temperatures of organisms in hot environments. The amount of cooling available to the organism depends on the concentrations of water vapor at the organism surface and in the environment, and on the conductance to water vapor of the organism surface and boundary layer. This chapter discusses terminology for specifying gas concentrations and provides information about concentrations of gases in terrestrial environments.

3.1 Specifying Gas Concentration

Concentrations of the main atmospheric constituents are often expressed as percentages or volume fractions. For mainly historical reasons, which have to do with methods of measurement, water vapor concentration is expressed in a number of different ways. For reasons that will become clearer as we get farther into the subject, there is a substantial advantage to expressing concentrations of all gases in terms of mole fraction (moles of substance per mole of air), and fluxes as moles per square meter per second. The relationship between density or concentration and amount of substance j in a gas is

$$\rho_j = \frac{n_j M_j}{V} \tag{3.1}$$

where n_j is the number of moles, V is the volume of gas, and M_j is the molecular mass. Since the mole fraction of j is the ratio of moles of gas j to moles of air:

$$C_j = \frac{n_j}{n_a} = \frac{M_a \rho_j}{M_j \rho_a}. \tag{3.2}$$

Here, M_a is the molecular mass of air and M_j is the molecular mass of component j. Table 3.1 gives molecular masses for the main constituents of the atmosphere.

The molar density, or ratio ρ_j/M_j, is the same for all gasses. At standard temperature and pressure (STP; $0°$ C and 101.3 kPa) the molar density of any gas is 44.6 mol m^{-3} (one mole of any gas occupies 22.4 liters). The molar density of gas will show up a lot in our equations, so we give it the special symbol $\hat{\rho}$. The variation of molar density with pressure and temperature is given by the Boyle–Charles law which states that the volume of a gas is inversely proportional to its pressure (p) and directly proportional to its Kelvin temperature (T). Using the Boyle–Charles law the molar density of air can be computed from:

$$\hat{\rho} = 44.6 \frac{p}{101.3} \frac{273.15}{T}. \tag{3.3}$$

TABLE 3.1.   Properties of the major constituents of air.

Gas	Molecular Mass (g/mol)	Mol fraction in air	Density at STP (kg m^{-3})
Nitrogen	28.01	0.78	1.250
Oxygen	32.00	0.21	1.429
Carbon dioxide	44.01	0.00034	1.977
Water vapor	18.02	0 to 0.07	0.804
Air	28.97	1.00	1.292

A middle range temperature for biophysical calculations is 293 K (20° C), giving $\hat{\rho} = 41.4$ mol m^{-3} at sea level (101.3 kPa).

The relationship between volume, temperature, and pressure for a perfect gas is

$$p_j V = n_j RT \tag{3.4}$$

where p_j is the partial pressure of gas j, and R is the gas constant, 8.3143 J mol^{-1} K^{-1}. Substituting Eq. (3.4) into Eq. (3.2) gives

$$C_j = \frac{p_j}{p_a} \tag{3.5}$$

so the mole fraction of a gas can be calculated as the ratio of its partial pressure and the total atmospheric pressure.

One more relationship between mole fraction and other measures is useful. If two gases with initial volumes V_1 and V_2 are mixed to make a volume V_a and the pressure is the same on all three volumes, then the volume fraction, V_1/V_a is equal to the mole fraction, n_1/n_a. Gas concentrations in air are often expressed as percentages, parts per million (ppm), or parts per billion (ppb) on a volume basis (volume of the pure gas divided by the volume of air). It can be seen here that these measures are directly related to the mole fraction.

One more version of the perfect gas law is also useful. The density of a gas is the molecular mass multiplied by the number of moles, and divided by the volume occupied by the gas (Eq. (3.1)). Substituting this into Eq. (3.4) gives the relationship between the partial pressure of a gas and its concentration:

$$p_j = \frac{\rho_j RT}{M_j}. \tag{3.6}$$

Example 3.1. In 1985 the average concentration of CO_2 in the atmosphere of the earth was estimated to be 344 ppm. What is the mole fraction, partial pressure, and density (concentration) of atmospheric CO_2 in air at 20° C?

Solution. Parts per million (ppm) means volumes of CO_2 in 10^6 volumes of air. Since the volume ratio is equal to the mole fraction, 344 ppm is the same as 3.44×10^{-4} moles/mole or 344μ mol/mol. Using Eq. (3.5), $p_c = C_c p_a$. If $p_a = 101$ kPa, then $p_c = 3.44 \times 10^{-4} \times 1.01 \times 10^5$ Pa $= 35$ Pa. For density Eq. (3.6) is used and rearranged to get

$$\rho_c = \frac{p_c M_c}{RT} = \frac{35 \text{ Pa} \times 44 \frac{\text{g}}{\text{mol}}}{8.31 \frac{\text{J}}{\text{mol K}} \times 293 \text{ K}} = 0.63 \frac{\text{g}}{\text{m}^3}.$$

To get the units to divide out, you may need to refer to Table 1.2. Note that a Pascal is a Newton per square meter, and that a joule is a Newton-meter. A Pascal is therefore equivalent to a joule per cubic meter.

3.2 Water Vapor: Saturation Conditions

If a container of pure water is uncovered in an evacuated, closed space, water will evaporate into the space above the liquid water. As water evaporates, the concentration of water molecules in the gas phase increases. Finally, an equilibrium is established when the number of molecules escaping from the liquid water equals the number being recaptured by the liquid. If the temperature of the liquid was increased, the random kinetic energy of the molecules would increase, and more water would escape. The equilibrium vapor pressure, established between liquid water and water vapor in a closed system is known as the saturation vapor pressure for the particular temperature of the system. The saturation vapor pressure is the highest pressure of water vapor that can exist in equilibrium with a plane, free water surface at a given temperature. The saturation vapor pressure is shown as a function of temperature in Fig. 3.1. Since the influence from any other gases present in the space above the water is negligible, the vapor pressure above the water surface is essentially the same whether the closed space is initially evacuated or contains air or other gases. The symbol e is used to represent the vapor pressure of water, and the saturation vapor pressure is denoted by $e_s(T)$, indicating that the saturation vapor pressure is determined by temperature. Tables giving saturation vapor pressure as a function of temperature can be found in List (1971) and in Table A.3 of the Appendix of this book. The mole fraction, which we frequently use in future computations, depends on both temperature and pressure. It is computed using Eq. (3.5), by dividing $e_s(T)$ by the atmospheric pressure. The main variable determining atmo-

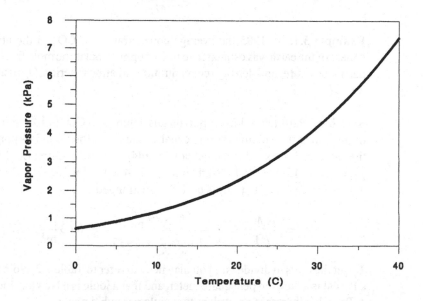

FIGURE 3-1. Saturation vapor pressure of air as a function of temperature.

spheric pressure is elevation. A relationship, which is accurate enough for biophysical calculations, is

$$p_a = 101.3 \exp\left(\frac{-A}{8200}\right) \tag{3.7}$$

where A is the altitude in meters above sea level and p_a is in kPa.

A convenient, empirical equation (with a close connection to the Clausius Clapeyron equation from thermodynamics) for computing the saturation vapor pressure from temperature is the Tetens formula (Buck, 1981):

$$e_s(T) = a \exp\left(\frac{bT}{T+c}\right) \tag{3.8}$$

where T is the Celsius temperature. The constants a, b, and c can be chosen to optimize the fit of the equation for various ranges of data. For environmental biophysics applications the constants are $a = 0.611$ kPa, $b = 17.502$, and $c = 240.97°$ C. Equation (3.8) can be used in place of tables for finding the saturation vapor pressure. While there are slight differences between the values from Eq. (3.8) and the more generally accepted values in List (1971), the differences are not measurable, nor are they significant for biophysical computations. The Tetens formula can also be used to predict the vapor pressure over ice (which is different from the vapor pressure over water). The coefficients for ice are $b = 21.87$ and $c = 265.5°$ C.

The slope of the saturation mole fraction with respect to temperature is also used frequently in computations. It is obtained by dividing the slope of the saturation vapor pressure function by atmospheric pressure. The slope of the saturation vapor pressure function is obtained by differentiating Eq. (3.8) to obtain

$$\Delta = \frac{bc\, e_s(T)}{(c+T)^2}. \tag{3.9}$$

The slope of the saturation mole fraction is represented by s, and is given by

$$s = \Delta/p_a. \tag{3.10}$$

Example 3.2. Find the saturation vapor pressure at 0, 10, 20, and 30° C, and the mole fraction of water vapor in saturated air at sea level for each of these temperatures.

Solution. When $T = 0$, $\exp(0) = 1$, so Eq. (3.8) gives $e_s(0) = a = 0.611$ kPa. The others require a little more computation, but, using Eq. (3.8) they give $e_s(10) = 1.23$ kPa, $e_s(20) = 2.34$ kPa, and $e_s(30) = 4.24$ kPa. Comparing these to the values in Table A.3 shows them to agree to the number of significant digits shown here. The pressure at sea level is 101 kPa, so the mole fraction at 0° C is

$C_v = 0.611$ kPa$/101$ kPa $= 0.006$. The other mole fractions are 0.012, 0.023, and 0.042 mol/mol or 6, 12, 23, and 42 mmol/mol.

There are a couple of points from the example that are worth noting. Vapor pressures and mole fractions are used in many calculations throughout this book, so it would be a good idea to find some way of remembering approximate values for these quantities at saturation. Note that the vapor pressure approximately doubles for each $10°$ C temperature increase. Exactly doubling would give 0.6, 1.2, 2.4, and 4.8 kPa for the four temperatures. All but the last one are within a few percent of the actual value, and it is only a little over ten percent high. If you can remember the vapor pressure at zero, you can therefore estimate saturation vapor pressures at higher temperatures. The other point to note is that conversion to mole fraction at sea level involves division by a number close to 100. The mole fraction, expressed as a percent, is therefore nearly the same as the vapor pressure. This is also the fraction of a saturated atmosphere made up of water vapor at the indicated temperature. The mole fraction, expressed in mmol/mol is just the vapor pressure in kPa multiplied by 10 (and is equal to the vapor pressure in millibars).

3.3 Condition of Partial Saturation

In nature, air is seldom saturated, so we need to know more than just the temperature to specify its moisture condition. Partial saturation can be expressed in terms of ambient vapor pressure or mole fraction, relative humidity, vapor deficit, dew point temperature, or wet bulb temperature. Ambient vapor pressure is simply the vapor pressure that exists in the air, as opposed to saturation vapor pressure, which is the maximum possible vapor pressure for the temperature of the air. Relative humidity is the ratio of ambient vapor pressure to saturation vapor pressure at air temperature:

$$h_r = \frac{e_a}{e_s(T_a)}. \tag{3.11}$$

Relative humidity is sometimes multiplied by 100 to express it as a percent rather than a fraction, but it is always expressed as a fraction in this book.

The relationship between saturation vapor pressure and ambient vapor pressure at various humidities is shown in Fig. 3.2. The curved lines, labeled on the right, show vapor pressures at humidity increments of 0.1 for temperatures from 0 to $40°$ C.

The vapor deficit is the difference in vapor pressure or mole fraction between saturated and ambient air:

$$D = e_s(T_a) - e_a = e_s(T_a)(1 - h_r) \tag{3.12}$$

where the second relation follows from Eq. (3.11). The vapor deficit at any temperature and relative humidity is the difference, in Fig. 3.2, between the saturation line ($h_r = 1$) at T_a and the line for the ambient relative humidity.

The dew point temperature is the temperature at which air, when cooled without changing its water content or pressure, just saturates. In other

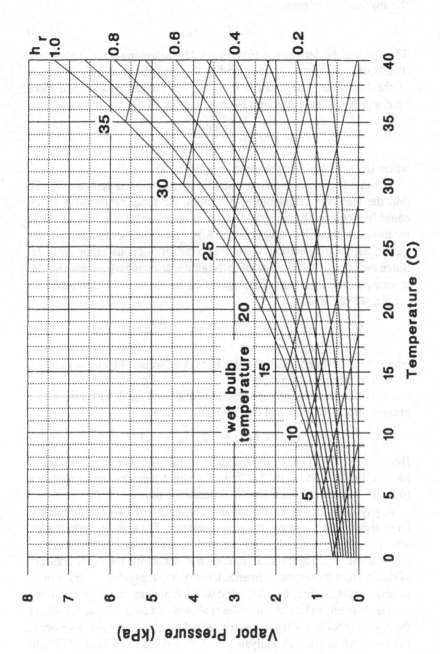

FIGURE 3.2. Vapor pressure-temperature-relative humidity-wet bulb temperature diagram. Wet bulb lines are for sea level pressure.

words, the saturation vapor pressure at dew point temperature is equal to the ambient vapor pressure:

$$e_s(T_d) = e_a. \tag{3.13}$$

This can also be determined from Fig. 3.2 by following horizontally from the ambient vapor pressure, and reading the temperature at the intersection of the horizontal line and the $h_r = 1$ line. It can be obtained more precisely from Table A.3, and by inverting Eq. (3.8):

$$T_d = \frac{c \ln(e_a/a)}{b - \ln(e_a/a)} \tag{3.14}$$

where the constants are the same as in Eq. (3.8).

Another important moisture variable is wet bulb temperature, T_w. To find the wet bulb temperature, determine the temperature drop which could be achieved by adiabatic evaporation of water into air (adiabatic meaning "without heat exchange"). Air is cooled by evaporating water into it, but the evaporation of water into the air raises its vapor pressure. Since the change in heat content of the air due to changing its temperature must equal the latent heat of evaporation for the water evaporated into the air, we can write:

$$c_p(T_a - T_w) = \lambda \left[C_s(T_w) - C_{va} \right] = \frac{\lambda \left[e_s(T_w) - e_a \right]}{p_a} \tag{3.15}$$

where λ is the latent heat of vaporization of water (44 kJ/mol) and c_p is the specific heat of air (29.3 J mol^{-1} K^{-1}). Equation (3.15) is most often written in terms of vapor pressure and used for determining vapor pressure from wet bulb and dry bulb temperatures:

$$e_a = e_s(T_w) - \gamma p_a(T_a - T_w). \tag{3.16}$$

Here, $\gamma = c_p/\lambda$ is called the thermodynamic psychrometer constant. It has a value of $6.66 \times 10^{-4} \, \text{C}^{-1}$ with a slight temperature dependence (0.01%/C) due to the temperature dependence of λ.

A psychrometer is an instrument consisting of two thermometers. One thermometer measures the air temperature. The second thermometer, whose "bulb" is covered with a wet cotton wick, measures the wet bulb temperature, T_w. Equation (3.16) is used to determine the vapor pressure of the air from these measurements. Clearly, a real psychrometer is not an adiabatic system since both heat and water vapor are exchanged with the surrounding air, and the thermometers absorb and emit radiation. Later in the book the tools needed to analyze a real psychrometer are developed, but the result of such an analysis yields an equation like Eq. (3.16) with an apparent psychrometer constant γ^*. For an adequately ventilated psychrometer with a good wick and radiation shield the value of the apparent psychrometer constant is close to the thermodynamic constant, but poorly designed or ventilated psychrometers can have much lower constants.

Equation (3.16) defines the family of straight, diagonal lines shown in Fig. 3.2. The wet bulb temperatures are labeled along the $h_r = 1$ line.

Either Eq. (3.16) or Fig. 3.2 can be used to find vapor pressure from wet and dry bulb temperatures, later we give examples of both. Finding wet bulb temperature from vapor pressure and air temperature using Fig. 3.2 is also easily done. Finding T_w from the psychrometer equation, however, is somewhat more challenging, since T_w appears in the linear term and implicitly in $e_s(T_w)$. There is no explicit solution to this equation, but a solution can be found by standard mathematical methods for solving nonlinear equations.

Example 3.3. When air temperature is $23°$ C and vapor pressure is 1.1 kPa, use Fig. 3.2 to find relative humidity, dew point temperature, and wet bulb temperature.

Solution. Enter Fig. 3.2 at $T = 23°$ C and $e_a = 1.1$ kPa. This point is just below the line for $h_r = 0.4$, so the humidity is estimated to be 0.39. The dew point temperature is the intersection of the $h_r = 1$ line and $e_a = 1.1$ kPa. The dew point temperature is therefore around $8°$ C. To find the wet bulb temperature, place a straight edge on Fig. 3.2 with the edge parallel to the wet bulb lines and passing through the point $T = 23°$ C, $e_a = 1.1$ kPa. Read the temperature on the scale along the $h_r = 1$ line, it is approximately $14.5°$ C. Note that the wet bulb temperature always lies between the air temperature and the dew point temperature unless the humidity is 1; then they are all equal.

Example 3.4. Find the vapor pressure and vapor mole fraction at the surface of melting snow at 1300 m elevation.

Solution. The temperature of melting ice and snow is $0°$ C. Since the surface is pure water, the vapor pressure at the surface is the saturation vapor pressure for that surface temperature. From Table A.3 the saturation vapor pressure at $0°$ C is 0.611 kPa. Using Eq. (3.7), the pressure at 1300 m elevation is $101.3 \exp(-1300/8200) = 86$ kPa. The vapor mole fraction is $C_v = 0.611$ kPa$/86$ kPa $= 0.007$ mol/mol or 7 mmol/mol.

Example 3.5. A psychrometer gives an air temperature of $30°$ C and a wet bulb temperature of $19°$ C. Use the psychrometer equation and other formulae to find the vapor pressure, relative humidity, dew point temperature, and vapor deficit of the air. Assume the measurement is made at sea level.

Solution. To use the psychrometer equation the saturation vapor pressure at wet bulb temperature needs to be known. Table A.3 gives $e_s(T_w) = 2.20$ kPa. The saturation vapor pressure at air temperature is also needed to find the humidity. Again, from Table A.3 it is $e_s(T_a) = 4.24$ kPa. Using the psychrometer equation (Eq. (3.16)) gives

$$e_a = 2.20\,\text{kPa} - 101\text{kPa} \times 6.67 \times 10^{-4}\text{C}^{-1}(30\,\text{C} - 19\text{C}) = 1.46\,\text{kPa}.$$

Now, using Eq. (3.11), $h_r = 1.46\,\text{kPa}/4.24\,\text{kPa} = 0.345$. The vapor deficit (Eq. (3.12)) is $D = 4.24\,\text{kPa} - 1.46\,\text{kPa} = 2.78\,\text{kPa}$. The dew point temperature can be obtained from Table A.3 or from Eq. (3.14). In Table A.3, follow down the vapor pressure column until you find the vapor pressure just smaller than 1.46 kPa. This is at 12° C. The dew point temperature therefore lies between 12° C and 13° C. The difference between the value at 12° C and 1.46 kPa is 58 Pa; the difference between the value at 12 and 13° C is 95 Pa. Our value of 1.46 kPa is therefore 58/95 or 0.6 of the way between 12 and 13° C. The dew point temperature is therefore $T_d = 12.6°$ C. Using Eq. (3.14):

$$T_d = \frac{240.97 \ln \frac{1.46}{0.611}}{17.502 - \ln \frac{1.46}{0.611}} = 12.62°\text{C}.$$

Example 3.6. The vapor pressure of air at a ski resort is 0.4 kPa. A machine to make artificial snow emits water droplets which quickly cool to the wet bulb temperature. At what air temperature will the droplets just reach freezing temperature at an altitude of 2400 m?

Solution. Ice melts at 0° C, but often cools a few degrees below this temperature before freezing (super cools). For this problem we will assume a freezing temperature of $-2°$ C. Rearranging the psychrometer equation (Eq. (3.16)) to find air temperature gives

$$T_a = T_w + \frac{e_s(T_w) - e_a}{\gamma p_a}$$

$$= -2 + \frac{0.53\text{kPa} - 0.4\text{kPa}}{6.67 \times 10^{-4}\text{C}^{-1} \times 0.75 \times 101\text{kPa}} = 0.6°\text{C}.$$

Therefore, expect the droplets to start freezing at an air temperature slightly above zero.

Example 3.7. A humidity sensor at air temperature reads 0.23 when the air temperature is 16° C. What is the vapor pressure, dew point temperature, and wet bulb temperature?

Solution. Entering Fig. 3.2 at the $T = 16°$ C line and $h_r = 0.23$ gives a vapor pressure of about 0.4 kPa. The wet bulb temperature is obtained by moving along a line through the vapor pressure, temperature point, parallel to the psychrometer lines. It is approximately 7.5° C. The dew point temperature cannot be obtained from Fig. 3.2 because it does not go low enough. Table A.3 also fails to go low enough for this problem. Before turning to the formula to obtain this value, try calculating a more precise value for the vapor pressure. From Table A.3, the saturation vapor pressure at 16° C (air temperature) is 1.82 kPa. Rearranging Eq. (3.11) to compute vapor pressure gives:

$$e_a = h_r e_s(T_a) = 0.23 \times 1.82\,\text{kPa} = 0.42\,\text{kPA}.$$

TABLE 3.2.   Comparison of various measures of moisture in
air for air temperature $= 20°$ C, relative humidity $= 0.5$, and
atmospheric pressure $= 100$ kPa.

$e_s(20)$ (kPa)	e_a (kPa)	T_d (C)	T_w (C)	C_{va} (mmol/mol)	ρ_v (gm^{-3})	r (g/kg)	q (g/kg)
2.34	1.17	9.3	14.0	11.7	8.65	7.36	7.31

Now, using Eq. (3.14) the dew point temperature is obtained:

$$T_d = \frac{240.97 \ln \frac{0.42}{0.611}}{17.502 - \ln \frac{0.42}{0.611}} = -5.1 \text{ C}.$$

Three other quantities are commonly used by meteorologists to describe vapor concentration in air. These are the absolute humidity, the
mixing ratio, and the specific humidity. The absolute humidity, also
known as the vapor density, is the mass of water vapor per unit volume of air. It is related to the vapor pressure by Eq. (3.6). For water vapor
this becomes

$$\rho_v = \frac{e_a M_w}{RT} \tag{3.17}$$

where e_a has units of Pa. If e_a has units of kPa, and ρ_v has units of
g m^{-3}, then for $\mathbf{T} = 293$ K, $\rho_v = 7.4e_a$.

The mixing ratio r is the mass of water vapor per unit mass of dry air.
It can be computed from the mole fraction of water vapor using

$$r = \frac{0.622 C_{va}}{1 - C_{va}}. \tag{3.18}$$

The specific humidity q is the mass of water vapor divided by the mass
of moist air, and is related to the mole fraction by:

$$q = \frac{0.622 C_{va}}{1 - 0.378 C_{va}}. \tag{3.19}$$

The units of r and q usually are g/kg. Table 3.2 compares values of r and
q with sample values of the other variables representing moisture in air.

3.4 Spatial and Temporal Variation of Atmospheric Water Vapor

The spatial and temporal patterns of vapor pressure in the atmosphere
resemble those given in Ch. 2 for temperature. During the day, vapor
pressures are highest near a soil or plant surface and decrease with height.
At night, vapor pressures tend to be lowest near the surface and increase
with height. Vapor pressures tend to be a bit higher in the day than at night
and typically reach a minimum at the time temperature is at the minimum.
As with temperature, the surface acts as a source of water vapor in the day

and a sink at night (when condensation and dew formation occur), and is therefore responsible for the shape of the vapor pressure profiles. Because the surface acts as a source or sink, and water vapor is transported in the atmosphere, there also exist high frequency random fluctuations in vapor pressure like those for temperature shown in Fig. 2.4.

While these patterns are easily demonstrated, the magnitude of spatial and temporal vapor pressure variation is much smaller than for temperature, and is usually small enough that it can be ignored in comparison with other sources of uncertainty in the measurements. If only the average vapor pressure for a day is known, the best estimate of hourly vapor pressures is that they equal the average for the day. Variation of vapor pressure with height can be described by an equation similar to Eq. (2.1), so a log plot of two or more measured vapor pressures with height would allow extrapolation or interpolation to other heights, as was done with temperature. However, the changes in vapor pressure with height are relatively small, so vapor pressures in an organism microenvironment are similar to the the vapor pressure at measurement height.

Not all measures of atmospheric moisture are as well behaved as vapor pressure or mole fraction, however. Figure 3.3 shows the diurnal variation in relative humidity and vapor deficit for the temperatures in Fig. 2.2, assuming the vapor pressure is constant throughout the day at 1.00 kPa ($T_d = 7°$ C). Note that the humidity is near one and the vapor deficit is near zero early in the morning. In the early afternoon the humidity is around 0.3 and the vapor deficit is 2 kPa. All of this variation is

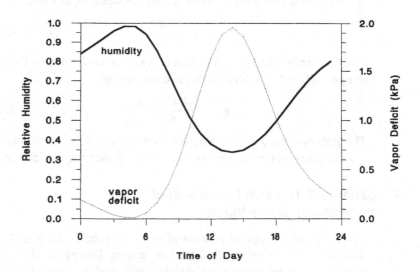

FIGURE 3-3. Diurnal variation in relative humidity and atmospheric vapor deficit for the temperature variation in Fig. 2.2. Vapor pressure is assumed to be constant throughout the day at 1.00 kPa.

brought about by the change in temperature, with no change in the vapor concentration in the air.

It is perhaps unfortunate that one of the most common measurements of atmospheric moisture is relative humidity. The measurement itself is essentially useless as an environmental variable except as a means, along with air temperature, of obtaining the vapor pressure, mole fraction, or dew point temperature. Some people compute and report averages of humidity over time periods of a day or longer. It should be clear from Fig. 3.3 that an average humidity is worse than meaningless. In addition to failing to communicate any useful information by itself, averaging individual humidity measurements destroys any possibility of obtaining useful information from the original data because the average humidity depends on the pattern of temperature variation (which is lost in the averaging process). It is best to immediately convert humidity data to vapor pressure or dew point. Then record, average, and process these data.

Averaging the vapor deficit is a slightly different matter. We show later that the vapor deficit gives an estimate of the driving force for evaporation, and is useful in relating transpiration and biomass production in plant communities. The average vapor deficit for the atmosphere is therefore a useful number, but it can be estimated reliably from average vapor pressure and temperature.

3.5 Estimating the Vapor Concentration in Air

Reliable measurements of atmospheric moisture are difficult to obtain, but estimates of the vapor pressure, which are quite reliable, are relatively easy to make. In the absence of airmass changes and advection, the vapor pressure in the air is relatively constant throughout the day and from day to day. It also varies little between indoors and outdoors. Figure 3.3 shows the humidity going to nearly 1.0 at the time of minimum temperature. This behavior is typical of all except arid, summer conditions. Therefore, the minimum daily temperature can often be taken as the dew point temperature.

Example 3.8. Summer minimum temperatures at locations in the Midwestern U.S. can be around 20° C, while in the arid Southwest they are 10° C or below. Compare the vapor pressure of the two locations.

Solution. Assuming the minimum temperature equals the dew point temperature, the vapor pressure can be looked up either in Fig. 3.2 or Table A.3. At 20° C it is 2.3 kPa and at 10° C it is 1.2 kPa. That difference makes an enormous difference in human comfort when temperatures are in the high 30s (C).

Example 3.9. On a particular foggy, cold day the outdoor temperature is $-20°$ C. Compare the humidity outdoors with the humidity in a heated building where air temperature is $22°$ C.

Solution. If there is fog, the humidity outdoors must be 1.0. In order to find the humidity indoors the vapor pressure of the air and the saturation vapor pressure at the indoor air temperature are needed. Assume that the indoor and outdoor vapor pressures are equal. Neither our figure nor our table give values at $-20°$ C, so we use Eq. (3.8):

$$e_a = 0.611 \exp \frac{17.502 \times (-20)}{240.97 - 20} = 0.125 \, \text{kPa}.$$

The saturation vapor pressure at $22°$ C is 2.64 kPa (Table A.3), so the indoor humidity is $h_r = 0.125/2.64 = 0.05$. Note that the same amount of vapor in the air gives very different humidities in the two environments.

Vapor pressure or humidity normally is measured in weather stations. Typically such measurements are taken 1.5 to 2 m above the ground in an open area. The humidity of microenvironments in plant canopies or near leaves can be quite different. For example, a tiny mite living on a corn leaf may be experiencing a relative humidity of 70 percent when the humidity measurement at an adjacent weather station indicates 30 percent. This can occur because the mite is small enough to be reside between the leaf surface and the top of the boundary layer surrounding the leaf, so that moisture from the transpiration stream leaving the stomata humidifies the mite environment. This humidification of leaf boundary layers is also important to transpiration because the conductance of the leaf surface (stomatal conductance) is influenced by the humidity in this leaf boundary layer. Humidity can vary greatly among various microenvironments, and we study such effects in following chapters.

References

Buck, A. L. (1981) New equations for computing vapor pressure and enhancement factor. J. Appl. Meteorol. 20:1527-1532.

Geiger, R. (1965) The Climate Near the Ground. Cambridge, Mass.: Harvard University Press.

List, R. J. (1971) Smithsonian Meteorological Tables, 6th. ed. Washington, D. C.: Smithsonian Institution Press.

Problems

3.1. A psychrometer gives air temperature of $34°$ C and wet bulb temperature of $22°$ C. Find the vapor pressure, the vapor mole fraction, the dew point temperature, the relative humidity, and the vapor deficit. The altitude of the site is 1200 m.

3.2. On a hot, humid day your skin surface can be at $36°$ C and covered with perspiration. If the perspiration has the same vapor pressure as pure water, what is the vapor pressure at your skin surface?

3.3. You decide to calibrate a relative humidity sensor at a value of 0.5 in a 0.1 m^3 container. Assuming no water is adsorbed on the walls of the container, how much liquid water must be evaporated into dry air to achieve this humidity at container temperatures of 10 and $40°$ C?

3.4. The method in problem 3.3 is not very useful for calibrating humidity sensors. A better method is to bubble air through water at controlled temperatures. Suppose air is bubbled through water at $10°$ C, and then passed into the container with the humidity sensor, which is at $20°$ C. What is the relative humidity of the air in the chamber? If the air and humidity sensor were really at $21°$ C when you thought they were at $20°$ C, how much error would this cause in your calibration?

3.5. If the outdoor minimum temperature were $-15°$ C on a particular day, estimate the relative humidity in a $22°$ C laboratory on that day.

3.6. The vapor pressure of the air you breathe out is the saturation vapor pressure at body temperature ($37°$ C). If the air you breathe in is at $20°$ C and 0.2 relative humidity, you take 15 breaths per minute, and each breath has a volume of 1 liter, how many 250 ml. glasses of water are required per day to replenish this water loss?

3.7. If air is at $15°$ C and 0.6 relative humidity, find the absolute humidity (or vapor density), the specific humidity, and the mixing ratio.

Liquid Water in Organisms and their Environment 4

Almost all of the water in living organisms is liquid, rather than vapor. In addition, water is taken up from the organism environment mainly in the liquid phase. Good physical descriptions of water in the liquid phase are necessary to understand liquid-phase water exchange and organism response. The energy state of liquid water can also affect the vapor pressure and concentration of water at evaporating surfaces. Vapor exchange is therefore also influenced by the state of the liquid water.

4.1 Water Potential and Water Content

Two types of variables are required to describe the state of matter or energy. One describes the amount, while the other describes the quality or intensity. For example, the thermal state of a substance is described in terms of its heat content and its temperature. While the two variables are related (the higher the temperature of a substance, the higher the heat content), they are not equivalent. The heat content depends on the mass, specific heat, and temperature of the substance, and gives no indication of which direction heat will flow if the object is placed in contact with another object. The temperature of an object specifies the intensity or quality of the heat, and temperature differences relate directly to direction and rate of heat flow.

Variables like temperature, which describe intensity of quality are called intensive variables, while variables describing amount are called extensive variables. Temperature is intensive while heat is extensive. To refer to thermal time as "heat units" is a confusion of intensive and extensive variables.

Describing the state of water in a system also requires the use of intensive and extensive variables. The extensive variable is familiar to most, and is called the water content. The intensive variable, called the water potential, is less familiar. Like temperature, it determines the direction and rate of water flow. As with temperature, there is often a relationship between the water content and the water potential of a substance, though it is always much more complicated than is the case for temperature. The important thing to realize is that water in soil, or in the tissues of living things, is not like the water in a glass. It is bound by the tissue or soil matrix, diluted by solutes, and sometimes is under pressure or tension. Its energy state is therefore quite different from that of water in a glass.

The water content is simply the ratio of the volume of water in a material to its total volume, or the ratio of mass of water to dry or wet mass of the material. Different bases are used as the standard in different disciplines, and all are called water content, so it is easy to make mistakes if one is not careful. In this book water content is defined as:

$$\theta = \frac{V_w}{V_t}$$
$$w = \frac{m_w}{m_d}$$

(4.1)

where V is the volume, m is the mass, and subscripts w, t, and d refer to water, total, and dry volume or mass. We call θ volumetric water content and w the mass water content. These are related by

$$\theta = \rho_b w / \rho_w$$

where ρ_b is the bulk density

$$\rho_b = \frac{m_d}{V_t}.$$

Water potential is defined as the potential energy per mole, per unit mass, per unit volume, or per unit weight, of water, with reference to pure water at zero potential. In thermodynamic terms, the energy per mole is the molar Gibbs free energy of the water in the system. A gradient of the water potential is the driving force for liquid water movement in a system.

As indicated, several sets of units are in use to describe water potential. For consistency with the rest of this book, we should use energy per mole, but this has not been used elsewhere, and may be completely unfamiliar to readers. Our preference is for energy per unit mass (J/kg). The units clearly show energy and mass, and, unlike volume, the mass does not vary with the density of the water. Energy per unit volume (J/m^3) is dimensionally equivalent to pressure (kPa or MPa). These units are frequently used for water potential, but fail to indicate a relationship to specific energy and have a less sound basis for the computation (the specific volume of water varies with density and is therefore dependent on temperature and binding energy). While these are minor objections to the use of pressure for water potential, it should be pointed out that there certainly are no advantages to the use of pressure units, and the mass-based units have historical priority. Energy per unit weight (J/N) is dimensionally equivalent to the height of a water column (m) in a gravitational field. It is used mainly in soil water flow problems where height of a physical water column is a convenient reference for other potentials. If the density of water is assumed to be 1 Mg/m^3, and the gravitational constant is 9.8 m s^{-2}, then

$$1 \text{ J/kg} = 1 \text{ kPa} = 0.001 \text{ MPa} \approx 0.1 \text{ m}.$$

In this book we use J/kg, but the reader can consider those equivalent to kPa if pressure units are more familiar.

The water potential is made up of several components. The total potential is usually written as the sum of the components:

$$\psi = \psi_g + \psi_m + \psi_p + \psi_o \qquad (4.2)$$

where the subscripts, g, m, p, and o are for gravitational, matric, pressure, and osmotic components. While each of these components (and others that could be defined) can contribute to the total potential, there are many situations where only one or two of the component potentials are active.

The gravitational potential is the potential energy of water as a result of its position in a gravitational field. A reference height must be specified in order to compute a gravitational potential. The gravitational potential is then:

$$\psi_g = gh \qquad (4.3)$$

where g is the gravitational constant (9.8 m s^{-2}) and h is the vertical distance from the reference height to the location where potential is specified. Above the reference h is positive and below the reference it is negative.

The matric potential arises from the attraction between water and soil particles, proteins, cellulose, etc. Adhesive and cohesive forces bind the water and reduce its potential energy compared to that of free water. For any substance that imbibes water there exists a relationship between water content and matric potential. This relationship is called the moisture characteristic. Figure 4.1 shows moisture characteristics for soils with three different textures. Most of the water in the clay is held very tightly (at low potential) because the large surface area of the clay is able to bind the water. Most of the water in the sand is held loosely (at high potential) because the sand matrix is ineffective in binding water. Similar curves could be made for cellulose, protein, etc., and they would have similar shapes. Tracy (1976) obtained a moisture characteristic for a whole frog. Note that the matric potential is always negative or zero. An empirical equation that closely approximates most moisture characteristics over a wide range of matric potentials is:

$$\psi_m = aw^{-b} \qquad (4.4)$$

where w is the water content and a and b are constants determined from data.

The pressure potential arises as a result of an applied hydrostatic or pneumatic pressure. Examples of this potential are the blood pressure in an animal, the water pressure under a water table in the soil, the turgor pressure inside plant cells, or the air pressure inside a pressure vessel which measures water potential in leaves or matric potential in soil. In many cases the pressure potential is hard to distinguish from the matric potential. For example, in soil a positive hydrostatic pressure is called a pressure potential and a negative pressure a matric potential. In the xylem of plants the pressures are generally negative, but the potential is referred

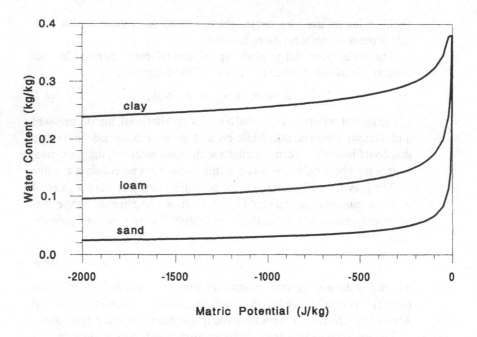

FIGURE 4.1. Soil moisture characteristics for three soil textures.

to as a pressure potential. Confusion arises because the components of the water potential differ in different systems, and because the components are defined primarily by method of measurement rather than on some rigorous thermodynamic basis. For our definitions, we attempt to distinguish between pressure and matric potential in terms of the nature of the forces acting on the water. Matric potential is defined as the reduction in water potential from short-range forces near interfaces (capillary forces or van der Waals forces). It is always negative. Pressure potential is considered a more macroscopic effect acting throughout a larger region of the system.

The pressure potential is computed from:

$$\psi_p = \frac{P}{\rho_w} \qquad (4.5)$$

where P is the pressure (Pa) and ρ_w is the density of water. The pressure potential can be either positive or negative, but usually is just positive.

The osmotic component arises from the dilution effect when solutes are dissolved in water. It does not really act as a potential or driving force for water movement unless the solutes are constrained by a semipermeable membrane. This occurs mainly in plant and animal cells and at air-water interfaces. When the solute is constrained by a perfect membrane, the osmotic potential can be computed from:

$$\psi_o = -C\phi\nu RT \qquad (4.6)$$

where C is the concentration of solute (mol/kg), ϕ is the osmotic coefficient, v is the number of ions per molecule (e.g., 2 for NaCl, 3 for $CaCl_2$, and 1 for sucrose), R is the gas constant ($8.3143 \, J \, mol^{-1} K^{-1}$), and T is the kelvin temperature. The osmotic coefficient has a value of one for an ideal solute, and is generally within ten percent of that value for solutions encountered in organisms and their environment. More accurate values are available in Robinson and Stokes (1965).

Two examples from nature illustrate the balance of potentials and the way they sum. In plant cells, concentrations of solutes are quite high. The cell membrane is permeable to water but not to the solutes, so water tends to move into the cell. The cell wall prevents volume expansion, so the pressure inside the cell increases. When the sum of the pressure and osmotic potential is equal to the water potential in the xylem, water ceases to move into the cell. If the cell walls of plants were not rigid and able to withstand high pressures, water would continue to move into the cell, diluting its contents until life processes would cease.

The other example has to do with blood in the circulatory system of animals. Solutes are free to diffuse through the walls of the capillary system, but proteins are too large and are kept in the blood stream. The negative matric potential of the blood proteins just balances the positive blood pressure potential. The blood matric potential (referred to as colloid osmotic pressure in the medical literature), provides just enough "suction" to keep the blood in the circulation system.

Example 4.1. If the reference for gravitational potential is the water table at 2 m depth, what is the gravitational potential at the soil surface?

Solution. Using Eq. (4.3), $\psi_g = 2 \, m \times 9.8 \, m \, s^{-2} = 19.6 \, m^2 \, s^{-2}$. From the example in Ch. 1, we know that this is equivalent to 19.6 J/kg.

Example 4.2. If the osmotic potential of plant sap is equivalent to 0.3 molal KCl, and the total water potential of the tissue is -700 J/kg, what is the turgor pressure?

Solution. Using Eq. (4.6) to get the osmotic potential, with $C = 0.3$ mol/kg, $\phi = 1$, and $v = 2$ gives:

$$\psi_o = -0.3 \, \frac{mol}{kg} \times 1 \times 2 \times 8.31 \, \frac{J}{mol \, K} \times 293 \, K = -1461 \, \frac{J}{kg}.$$

Now use Eq. (4.2) to obtain the turgor pressure. Assume that all components except the osmotic and pressure components are negligible: $\psi_p = \psi - \psi_o = -700 \, J/kg - (-1461 \, J/kg) = 761 \, J/kg$. Using Eq. (4.5), $P = 761 \, J/kg \times 1000 \, kg/m^3 = 761 \, kPa$. One atmosphere is 101 kPa, so the pressure inside the cell is 7.5 atmospheres. If the plant were fully turgid (water potential of the leaf equal to zero), the pressure would be almost twice this value. These are typical values for plant leaves and illustrate the amazingly high pressures that routinely exist in living systems.

4.2 Water Potentials in Organisms and their Surroundings

It is useful for future computations to have some feeling for the range of water potentials that exist in organisms and their environment. Human blood has an osmotic potential around -700 J/kg. Fresh sweat is about half this concentration and urine is two to three times as concentrated as blood. Osmotic potentials of blood and other body fluids of most mammals are similar to these. The osmotic potentials of cell sap in plant leaves ranges from -500 to -7000 J/kg. Typical values for mesophytic species are in the range -1000 to -2000 J/kg. The water potential of leaves approaches that of the soil at night when transpiration rates are very low. If the soil is wet, the maximum leaf water potential is near zero. In the day, with high transpiration rates, the turgor pressure is close to zero, and the leaf water potential is about equal to the osmotic potential. The variation in leaf water potential, for a plant growing in wet soil, may therefore vary from -100 to -2000 J/kg over a diurnal cycle.

When soils are saturated, their water potential is near zero, but gravity quickly drains them to potentials between -10 and -30 J/kg. The water content corresponding to this water potential is called field capacity. It is an approximate, but useful upper limit for available water in soil. As plants extract water from the soil, the water potential decreases until all remaining water is so tightly held that root water potentials cannot drop low enough to withdraw additional water. The water content below which minimal water extraction by plant roots occurs is called the permanent wilting point, and it corresponds roughly to the water content when soil has a water potential of -1500 J/kg. Again, this point is not exact, but sets a useful lower limit for water available to the plant. Soil near the surface is further dried by the air, and may reach potentials of -3×10^5 J/kg, but this drying only affects the top few decimeters of soil. The remainder of the soil profile is not likely to dry below about -2000 J/kg.

At the lower limit of water potential for living systems, some fungi are able to live at water potentials in the range -50 to -70 kJ/kg, and there are reports of both plants and insects taking up water from environments which are this dry (though their internal water potentials are probably much higher). These, however, are very unusual situations.

With these water potentials in mind, we now consider the effect of water potential on the vapor pressure at the liquid–vapor interface.

4.3 Relation of Liquid- to Gas-Phase Water

For every computation of evaporation rate the vapor concentration at the evaporating surface needs to be known. This surface is the interface between the liquid water phase and the gas phase. In Ch. 3 we said that the concentration of vapor at this interface is the saturation vapor concentration at surface temperature if the surface is free water. If the surface is not a free water surface, then we expect the surface to have a humidity less than 1.0 and a vapor concentration less than the saturation concentration. From Eq. (3.11), we can write:

$$C_{vs} = h_{rs} \frac{e_s(T_s)}{p_a} = h_{rs} C_v(T_s) \tag{4.7}$$

where h_{rs} is the humidity at the liquid-gas interface. We expect this humidity to be related to the water potential of the liquid phase, but need to find the relationship.

The relationship between water potential and humidity can be derived by considering the work required to create a volume dV of water vapor. The first law of thermodynamics states that the change in internal energy (U) of a system is equal to the difference between the heat input (Q) and the work done. Restricting the work to volume expansion against an imposed pressure, then

$$dU = dQ - p \, dV. \tag{4.8}$$

If the system is adiabatic (no heat exchange) then $dQ = 0$. An expression for dV can be obtained by differentiating Eq. (3.4) to get

$$dv = -\frac{nRT}{p^2} \, dp. \tag{4.9}$$

Substituting Eq. (4.9) for dV in Eq. (4.8) gives

$$dU = \frac{nRT}{p} \, dp. \tag{4.10}$$

The change in energy in going from the reference state where $p = e_s$, the saturation vapor pressure, to $p = e$, some lower vapor pressure is obtained by integrating Eq. (4.10)

$$U = nRT \int_{e_s}^{e} \frac{dp}{p} = nRT \ln\left(\frac{e}{e_s}\right). \tag{4.11}$$

By Eq. (3.11), $h_r = e/e_s$. Also, $\psi = $ energy/mass $= U/nM_w$, where M_w is the molecular mass of water (0.018 kg/mol). Substituting these into Eq. (4.11) gives

$$\psi = \frac{RT}{M_w} \ln h_r. \tag{4.12}$$

The inverse relationship is more useful. It is

$$h_r = \exp \frac{M_w \psi}{RT}. \tag{4.13}$$

Example 4.3. Make a table of humidities at liquid–air interfaces for typical water potentials in organisms and their surroundings.

Solution. Assume a temperature of 293 K (20° C), then Eq. (4.13) becomes

$$h_r = \exp \frac{0.018 \frac{kg}{mol} \times \psi \frac{J}{kg}}{8.31 \frac{J}{mol\,K} \times 293\,K} = \exp \frac{\psi}{135000}.$$

Using this, the following table can be constructed.

Organism or location	Water Potential (J/kg)	Humidity
Blood	−700	0.995
Leaf, night	−100	0.999
Leaf, day; stressed	−2000	0.985
Xerophytic leaf	−8000	0.942
Xerophytic fungi	−58000	0.65
Soil, field capacity	−30	0.9998
Soil, permanent wilt	−1500	0.989
Soil, air dry	−100000	0.48
Saturated NaCl	−38000	0.755

The table shows that soils wet enough to support plant growth have humidities very near 1.0. The evaporating surfaces of animals and most leaves also have humidities near 1.0. Referring back to Eq. (4.7), it can generally be assumed that the vapor concentration at an evaporating surface is equal to the saturation vapor concentration at surface temperature. Departures from this occur when there are high concentrations of solute in the water, or when the surface has dried below water potentials typical of biological activity.

Example 4.4. In hot, arid environments, sweat evaporates quickly and leaves salts on the skin surface. Even though the concentration of salt in sweat is lower than that in blood, the concentration can eventually build up so that evaporation is finally occurring from a nearly saturated NaCl solution. Compare the vapor concentration at the evaporating surface just after a shower when the salt concentration is negligible with the concentration after a day of heavy work in the heat so that the skin is covered with salt. Assume the skin temperature is 36° C.

Solution. From Table A.3, the saturation vapor pressure at 36° C is 5.9 kPa. At sea level this is equivalent to a concentration of 59 mmol/mol. Just after a shower, $h_{rs} \approx 1$, so the concentration at the evaporating surface is 59 mmol/mol. When the skin is covered with salt, the humidity at the evaporating surface is around 0.75 (from the table in Example 4.3) so the vapor concentration at the surface is $0.75 \times 59 = 44$ mmol/mol. If the vapor concentration in the air were 20 mmol/mol, the difference would be 39 mmol/mol in the first case and 24 mmol/mol in the second.

The presence of the salt would therefore reduce the rate of cooling by about 40 percent.

References

Robinson, R. A. and R. H. Stokes. (1965) Electrolyte solutions. Butterworths. London.

Tracy, C. R. (1976) A model of the dynamic exchanges of water and energy between a terrestrial amphibian and its environment. Ecological Monographs 43:293–326.

Problems

4.1. What is the gravitational potential at the top of a 30 m tall tree if the reference is the soil surface?

4.2. What is the matric potential of a loam soil at 0.15 kg/kg water content?

4.3. The brine in the Great Salt Lake is mostly sodium chloride, and reaches concentrations around 6 mol/kg in some places. Find the osmotic potential of the brine. The osmotic coefficient (ϕ) of 6 molal NaCl is 1.27.

4.4. Find the vapor pressure and vapor mole fraction at the evaporating surface of the Great Salt Lake (see 4.3) when its surface temperature is 18° C. Its elevation is 1280 m.

4.5. If the temperature of a leaf is 33° C, estimate the vapor pressure at the evaporating surfaces inside the stomata of the leaf.

4.6. If the total water potential of a leaf is −700 J/kg and the osmotic potential is −1200 J/kg, what is the turgor pressure in the cells?

4.7. Estimate the humidity inside an animal burrow where plants are observed to be growing.

4.8. In plants sucrose is actively pumped, using metabolic energy, into the phloem near the source of carbon fixation by photosynthesis. If the concentration of sucrose outside the phloem is 0.5 mol/kg, the concentration inside the phloem is 1 mol/kg, and the sieve tube elements are perfectly semipermeable, how much pressure could be built up inside the sieve tube elements (assume $\psi_{plant} = 0$)? If the sucrose is unloaded at some downstream location by the same mechanism, so that the sucrose gradient is just reversed from that at the loading site, what pressure difference will be maintained between the loading and unloading zones in the phloem? This pressure difference is the driving force for flow in the phloem.

Wind 5

As living organisms, we are most acutely aware of three things about the wind. We know that it exerts a force on us and other objects against which it blows, it is effective in transporting heat from us, and it is highly variable in space and time. A fourth property of the wind, less obvious to the casual observer, but essential to terrestrial life as we know it, is its effective mixing of the atmospheric boundary layer of the earth. This can be illustrated by a simple example. On a hot summer day about 10 kilograms (550 moles) of water can be evaporated into the atmosphere from each square meter of vegetated ground surface. This amount of water would increase the vapor concentration in a 100 m thick air layer by 100 $g \, m^{-3}$ (136 mmol/mol) if there were no transport out of this layer or condensation within it. This is much more water than the air could hold at normal temperature. The observed increase in vapor concentration in the first 100 meters of the atmosphere is typically less than 1 $g \, m^{-3}$, so we can see how effective the atmosphere is for transporting and mixing. A similar calculation (Monteith, 1973) shows that photosynthesis in a normally growing crop would use all of the CO_2 in a 30 m air layer above a crop in a day, yet measured CO_2 concentrations have diurnal fluctuations of 15 percent or less. Without the vertical turbulent transport of heat, water vapor, CO_2, oxygen, and other atmospheric constituents, the microenvironment we live in would be very inhospitable.

The influence of the surface on the atmosphere of the earth can extend from hundreds of meters at night to several thousand meters during days when surface heating is strong. This depth of influence of the surface on the atmosphere is called the planetary boundary layer. Through the depth of this planetary boundary layer, like all boundary layers that form between moving fluids and stationary surfaces, fluxes of momentum, heat, and mass decrease with height. The lowest 50 m of this planetary boundary layer is referred to as the surface layer; this is the region of most interest. In this region fluxes of momentum, heat, and mass are virtually constant with height and profiles of wind speed, temperature, and concentration are logarithmic.

In order to determine the force of the wind on, or the rate of heat transfer from living organisms in their microenvironments, it is necessary to know the wind speed in the vicinity of the organism. This requires an understanding of the behavior of average wind in the surface boundary layer of the earth. The behavior of the wind, in turn, is dictated by rates of turbulent transfer in the surface boundary layer. Turbulent transfer theory allows us to derive equations for wind, temperature, vapor density, and CO_2 profiles and fluxes, and are helpful later when we discuss plant canopies and their environment. In this chapter we first discuss the behavior and characteristics of the wind in natural, outdoor environments. In a later chapter we present some of the fundamentals of turbulent transport theory and derive the profile equations for wind, temperature, and vapor concentration.

5.1 Characteristics of Atmospheric Turbulence

As was previously mentioned, one of the obvious characteristics of wind is its variability. We are aware of random temporal variations of the wind through fluttering of flags and leaves, variations in the force of wind on us, and other common experiences. Spatial variations are obvious when one looks at a field of "waving grain" or at "cat's paws" on a lake. We are also aware that the range of variability is large. We see very small scale fluctuations in "heat waves" on hot summer days and feel or hear the effects of very large scale fluctuations as wind gusts which blow dust or shake the house. All of these characteristics of wind with which we are intimately acquainted are characteristics of turbulent flow. Except for a thin layer of air close to surfaces, the atmosphere is essentially always turbulent, or, in other words, characterized by random fluctuations in wind speed and direction caused by a swirling or eddy motion of the air. These swirls or eddies are generated in two ways. As wind moves over natural surfaces, the friction with the surface generates turbulence. This is called mechanical turbulence. Turbulence is also generated when air is heated at a surface and moves upward due to buoyancy. This is called thermal or convective turbulence. The size of the eddies produced by these two processes is different, as is shown in Fig. 5.1. The fluctuations from mechanical turbulence tend to be smaller and more rapid than thermal fluctuations. A striking demonstration of these types of turbulence can be seen by watching the plume from a smokestack on a hot day. The plume is called a looping plume because, in addition to the small scale mechanical turbulence that tears the plume apart and spreads it with distance, the thermal updrafts and downdrafts cause the entire plume to be transported upward or downward.

Large eddies, which are produced either mechanically or thermally, are unstable and decay into smaller and smaller eddies until they are so small that viscous damping by molecular interactions within the eddies finally turns their energy into heat. The size of the smallest eddies produced by mechanical and convective motion (rather than breakdown of larger eddies) is called the outer scale of turbulence. The eddy size at which significant molecular interaction (viscous dissipation) begins is called

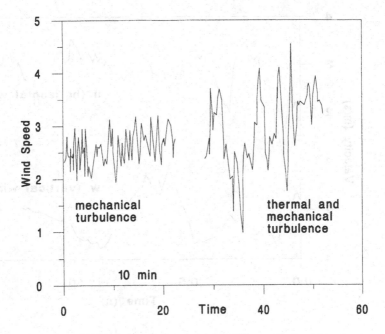

FigURE 5-1. Typical traces of a fast response wind sensor in conditions of pure mechanical turbulence and thermal plus mechanical turbulence.

the inner scale of turbulence. The outer scale is generally taken to be a few meters and the inner scale a few millimeters.

This process of larger eddies breaking down into ever smaller ones is expressed in a rhyme by L.F. Richardson, a scientist who is responsible for much of the theory of atmospheric turbulence (Gifford, 1968):

> Great whirls have little whirls
>
> That feed on their velocity;
>
> And little whirls have lesser whirls,
>
> And so on to viscosity.

Gifford remarks that this parody of de Morgan's verse on fleas "may be the only statement of a fundamental physical principle in doggerel."

5.2 Wind as a Vector

There is a fundamental difference between wind and the other environmental variables (temperature, vapor, and other gas concentrations) discussed in earlier chapters. Wind is a vector quantity involving both magnitude and direction, while the other environmental variables are scalars, where only magnitude is specified. The wind velocity vector is commonly divided into components along the axes of a rectangular coordinate system. For convenience, the coordinate system is oriented so that the x axis points in the direction of the mean wind. The velocity

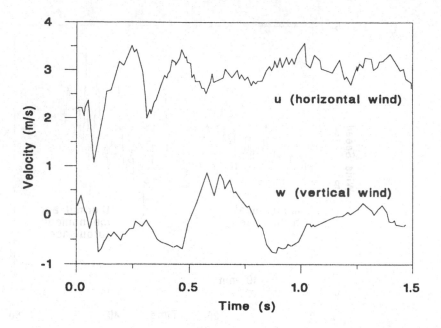

FIGURE 5-2. Fluctuations in horizontal and vertical wind recorded with fast response sensors (after Tatarski, 1961).

components are given the symbols u, v, and w, where u is the velocity component in the x direction, v in the y direction, and w in the z (vertical) direction. Each velocity component has a mean value (averaged over a period ranging from 15 minutes to an hour) and a component which fluctuates about the mean. Since the coordinate system is oriented in the direction of the mean wind, the mean v and w components are zero. Figure 5.2 shows the u and w components of the wind vector for a period of a few seconds. These can be compared with the temperature fluctuations in Fig. 2.4. Wind fluctuation (turbulence) is the underlying cause of all of these fluctuations, and similar patterns are found in measurements of all atmospheric scalars (water vapor, CO_2, etc.)

5.3 Modeling the Variation in Wind Speed

We return to turbulent transport under the topic of transport processes. We now focus on modeling just the variation in the mean wind speed. Our goal is the ability to model the wind speed around animals and leaves. These models are used later to compute the boundary layer conductances of these objects. We are therefore not so concerned with the fluctuations or vector components of the wind and are most interested in the wind speed. A rigorous analysis would distinguish between the wind speed and the magnitude of the mean wind velocity vector, but in practice the two are often used interchangeably. We therefore use the symbol u to represent both the magnitude of the velocity vector and the wind speed.

If average wind speed were measured at several heights above a soil surface, the profile would look like the lower ($h = 1$ cm) profile plotted in Fig. 5.3. The wind speed is zero at the soil surface, increases rapidly with height near the surface, and is fairly constant with height far from the surface. Over a crop (Fig. 5.3; $h = 50$ cm) the shape of the profile is similar, but is shifted upward. For the same height far from the surface, wind speeds over the crop are lower, and the wind profile appears to extrapolate to zero somewhere above the soil surface (though the actual wind speed in the crop is not zero, as is shown later).

The shape of these profiles is similar to the shape of the temperature profiles presented in Ch. 2. The reason for the shape is the same as for the temperature profiles. Turbulent transport near the ground is inefficient, with only small eddies forming to transport the heat or momentum. Farther from the surface the eddies are larger and the transport more efficient, so the gradients of temperature or wind speed are smaller.

As with temperature, we could log-transform the height axis of Fig 5.3 to obtain approximately linear relationships with wind speed. The offset

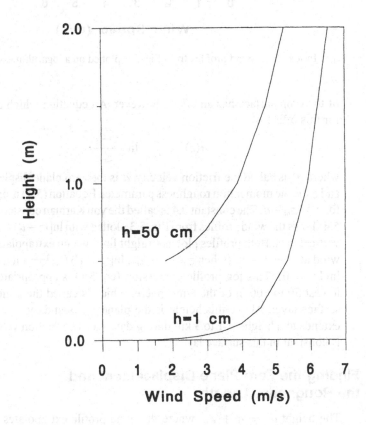

FIGURE 5-3. Profiles of wind speed above a 50 cm tall crop and a soil surface ($h = 1$ cm).

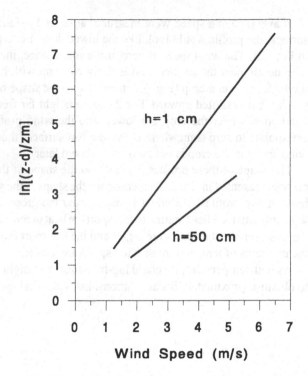

FIGURE 5.4. Wind profiles from Fig. 5.3 plotted on a logarithmic scale.

of the crop surface has an effect, however. An equation which accounts
for this offset is

$$u(z) = \frac{u^*}{0.4} \ln \frac{z - d}{z_m} \qquad (5.1)$$

where u^* is called the friction velocity, d is the zero plane displacement,
and z_m is the momentum roughness parameter. Equation (5.1) is only valid
for $z \geq z_m + d$. The constant 0.4 is called the von Karman constant. Figure
5.4 shows the wind profiles from Fig. 5.3 plotted with $\ln[(z - d)/z_m]$ as the
vertical axis. Both profiles plot as straight lines which extrapolate to zero
wind at $z = d + z_m$ (when $z = d + z_m$, $\ln[(z - d)/z_m] = \ln(z_m/z_m) =$
$\ln(1) = 0$). This log-profile expression (eq. 5.1) is appropriate for the
lowest 50 to 100 m of the atmosphere, which is called the atmospheric
surface layer. Above this height is the planetary boundary layer, which
extends to a height of 1 to 3 km during daytime. The friction velocity, u^*
is constant in this surface layer.

5.4 Finding the Zero Plane Displacement and the Roughness Length

The height $d' = d + z_m$, where the wind profile extrapolates to zero,
is an actual measurable height within the canopy. Its meaning can be
understood as follows. For smooth surfaces, the drag of the surface on

the wind occurs at the surface, and $d' = 0$. For a crop, the drag occurs throughout the crop, but a single surface placed at some height below the top of the crop can be imagined which would have an effect equivalent to the wind. If z in Eq. (5.1) is measured from that point, then there would be no need for a zero plane displacement in the equation. Normally, however, we want to measure z from the soil surface. When doing this, d' needs to be subtracted to place the effective location of drag near the top of the crop.

An intuitive feeling for the meaning of z_m and u^* is harder to obtain. The units of u^* are velocity or speed. Its value is directly proportional to the wind speed at height z, as can be seen from Eq. (5.1), but also depends on the friction of the wind with the surface. Thus the name "friction velocity." While z_m has units of length, one should not try to interpret it as a measurable physical length. It is a measure of the form drag and skin friction in the layer of air that interacts with the surface. It is most reliably determined empirically by measuring wind speed at several heights above a surface and plotting $\ln(z - d)$ versus $u(z)$. A straight line should result which can be extrapolated to $u = 0$. If we set $u(z) = 0$ in Eq. (5.1), it can be seen that $\ln(z - d) = \ln(z_m)$. The intercept, where $u = 0$, is $\ln(z_m)$. The value of z_m is the exponential of this intercept. Table 5.1 gives several values of z_m determined in this way. It should be pointed out that these are values from particular experiments and are not necessarily representative of all surfaces like the one described. The wind can make the surface rougher or smoother and the direction of the wind with respect to rows or other regular features of the surface can have a big effect on the roughness length. The wind obviously has a big effect on the roughness of water surfaces, but the effect on plant canopies can also be substantial. Maki (1975) did an extensive set of z_m and d measurements on a full-cover Teosinte canopy while its height (h) changed from 0.68 m to 1.45 m (LAI increased from 2 to 6) and observed a strong linear relationship between z_m and u^* as well as d and u^* over a range of u^* from 0.05 to 0.5 m/s; furthermore, z_m was related to d. Fitting data from the five measurement

TABLE 5.1. Empirically determined values of roughness length for various surfaces (from Hansen, 1993).

Type of Surface	z_m (cm)	Type of Surface	z_m (cm)
Ice	0.001	Coniferous forest	110
Dry lake bed	0.003	Alfalfa	3
Calm open sea	0.01	Potatoes, 60 cm high	4
Desert, smooth	0.03		
Grass, closely mowed	0.1		
Farmland, snow covered	0.2	Cotton, 1.3 m tall	13
Bare soil, tilled	0.2-0.6	Citrus orchard	31-40
Thick grass, 50 cm high	9	Villages, towns	40-50
Forest, level topography	70-120	Residential, low density	110
		Urban bldgs, business dist.	175-320

dates results in $z_m/h = 0.16u^*$ and $1 - d/h = 0.4u^*$ (u^* is in units of m/s here); this means that $1 - d/h = 2.5z_m/h$. The dependence of z_m and d on wind speed is rarely taken into account because it adds a complication that is not well understood.

For uniform vegetated surfaces, such as agricultural crops, the zero plane displacement and roughness length can be approximated by knowing just the height of the canopy. These relationships were worked out empirically several decades ago, but were improved upon by Shaw and Pereira (1982) using a simulation model of canopy–atmosphere interaction. By using the model they were able to investigate effects of plant density and foliage distribution with height in the canopy. Figure 5.5 shows the ratios d/h and z_m/h for a canopy with foliage distribution similar to a corn canopy. In the figure, h is the height of the canopy and plant area index (PAI) is the area of leaves and stems per unit ground area (related to plant density).

Figure 5.5 shows a generally increasing zero plane displacement with plant density and an increasing and then decreasing roughness length. This behavior is about what we would expect. At low density momentum exchange occurs throughout the canopy, but with increasing density the wind is less able to penetrate and the exchange is forced higher and higher in the canopy. At very low plant density, increasing the density increases the roughness of the surface, but after PAI of about 0.6 the increasing density tends to smooth the surface. At PAI = 0, d and z_m should equal

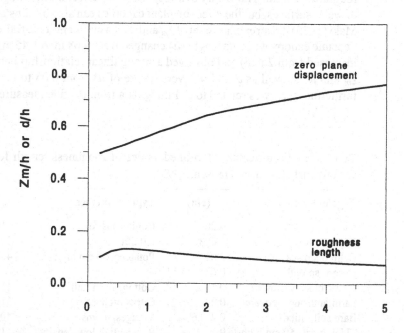

FIGURE 5.5. Change in d/h and z_m/h with plant area index for uniform plant canopies.

the values for the underlying surface. Figure 5.5 does not show this. However, the simulation was not intended to show the effects of very low plant density and the extrapolation to PAI = 0 is therefore not reliable.

If we take PAI = 3 as representative of a range of agricultural crops then Fig. 5.5 gives

$$d = 0.65h \qquad (5.2)$$

and

$$z_m = 0.1h. \qquad (5.3)$$

Using the simple relation between z_m and d from Maki (1975), if $z_m = 0.1h$ then $d = 0.75h$; this is in reasonable agreement with Fig. 5.5 at high PAI, where the relation of Maki (1975) would be expected to hold. When only canopy height is known, Eqs. (5.2) and (5.3) can be used to estimate d and z_m. If an estimate of the plant area density is available, then Fig. 5.5 can be used. If wind profile data are available, it is best to find z_m from the wind profile data.

Example 5.1. Four anemometers are to be placed at different heights above a plant canopy to measure wind speeds for a wind profile. At what heights should the anemometers be placed to give the most information about the wind profile? Assume the canopy is 40 cm tall.

Solution. Since wind speed varies with the logarithm of height above the zero plane, the most information is obtained (largest differences between readings) if the anemometers are logarithmically spaced. If the top anemometer is placed 2 m above $d + z_m$, then the others could be at 1, 0.5, and 0.25 m above $d + z_m$. The zero plane displacement is at $0.65 \times 0.4 = 0.26$ m, so the heights above the ground would be 0.51, 0.76, 1.26, and 2.26 m.

Example 5.2. The average wind speed measured 3 m above a golf course was found to be 2.7 m/s. What is the wind speed at the surface of the turf? Assume the grass is 3 cm tall.

Solution. From Eqs. (5.2) and (5.3), $d = 0.02$ m and $z_m = 0.003$ m. Using these, with $z = 3$ m gives

$$u(3) = 2.7 \text{ m/s} = \frac{u^*}{0.4} \ln \frac{3 - 0.02}{0.003}.$$

Solving for u^* gives $u^* = 0.16$ m/s. The friction velocity is constant for all heights, so this value can now be used to determine the average wind speed at any height. The wind speed at h (the top of the canopy) is therefore given by

$$u(h) = \frac{u^*}{0.4} \ln \frac{h(1 - 0.65)}{0.1h} = 3.13u^*.$$

Using the value of u^* just calculated, $u(0.03) = 3.13 \times 0.16 \, \text{m/s} = 0.5 \, \text{m/s}$.

5.5 Wind Within Crop Canopies

The equations given so far deal only with wind above the plant canopy. These equations are useful for determining the wind in the microenvironment of many living organisms, but cannot be used for leaves within a canopy or for animals or insects that live within a crop or forest canopy. Equation (5.1) predicts that wind speed is zero at $z = d + z_m$. It gives no information about wind speeds below this level, but presumably they would remain at zero to the bottom of the canopy. From experience, however, we know that the wind does blow inside canopies, so Eq. (5.1) is apparently wrong. In fact, Eq. (5.1) is only useful above the canopy. Within the canopy a different model must be used.

In order to model the wind in canopies, the canopy is divided into at least two layers. In most of the canopy the wind speed decreases exponentially with depth. Figure 5.6 shows the lower part of the wind profile from Fig. 5.3 for the 50 cm tall canopy along with the wind profile throughout most of the canopy. The equation for the top 90 percent of the canopy is:

$$u(z) = u(h) \exp\left[a\left(\frac{z}{h} - 1\right)\right] \tag{5.4}$$

where a is an attenuation coefficient for the crop. The wind speed at the

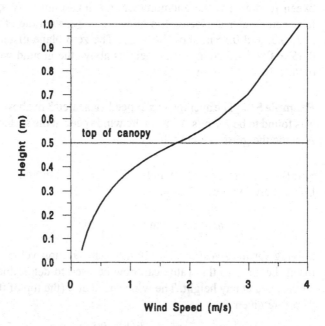

FIGURE 5.6. Wind speed within and just above a crop.

top of the canopy $u(h)$ is equal to $u(h)$ computed from Eq. (5.1), thus matching the two profiles at the top of the canopy. The value for a in Fig. 5.6 is 2.5. Table 5.2 gives values of a for a number of different canopies.

Goudriaan (1977) suggests a simple equation for calculating the attenuation coefficient as a function of crop structure; namely,

$$a \simeq \left(\frac{0.2 L_t h}{1_m} \right)^{1/2} \tag{5.5}$$

where L_t is the leaf area index, h is the canopy height, and 1_m is a mean distance between leaves in the canopy given by

$$1_m = \left(\frac{4wh}{\pi L_t} \right)^{1/2} \tag{5.6}$$

for grass leaves and

$$1_m = \left(\frac{6w^2 h}{\pi L_t} \right)^{1/3} \tag{5.7}$$

for leaves that are shaped more like squares; w is the leaf width. Table 5.3 contains some values of a calculated from Eq. (5.5) and compared to measurements. Clearly, Eq. (5.5) does not work well if the vegetation cover is too low such as in the first corn entry.

In the bottom 10 percent of the canopy, a new logarithmic profile is developed with a zero plane displacement of zero and a roughness length characteristic of the underlying soil surface. Equation (5.1) can therefore be used for this part of the canopy. The wind speed at the top of this layer is equal to the wind speed at the bottom of the exponential layer, so from one wind speed above the canopy all of the wind speeds can be estimated to the bottom of the canopy.

In tall tree canopies with dense foliage at the top and a relatively open stem space, the wind in the canopy can be quite unrelated to the wind above the canopy, in both speed and direction. An example of the behavior of the wind in this intermediate layer can be observed by watching the drift of smoke from a campfire in a forest. This wind results from horizontal pressure differences within the canopy, and is attenuated by drag of the elements within the stem space and by the ground surface. In

TABLE 5.2.   Attenuation coefficients for different crops (from Cionco, 1972)

Canopy	a	Canopy	a
Immature corn	2.8	Sunflower	1.3
Oats	2.8	Xmas trees	1.1
Wheat	2.5	Larch trees	1.0
Corn	2.0	Citrus orchard	0.4

TABLE 5.3. Canopy wind profile attenuation coefficient calculated for several canopies and compared to measurements in corn. Corn* is from Inoue and Uchijima (1979) and corn+ is from Sauer (1993). The values of w are estimated.

Crop	h (m)	LAI	w (m)	a (eq. 5.5)	a(meas)
Corn*	0.5	0.55	0.05	0.48	1.6
	1.4	2.5	0.08	1.7	2.0
	2.25	4.3	0.10	2.7	2.6
	2.77	4.2	0.12	2.7	3.0
Corn+	1.9	3.1	0.1	2.1	1.8
	2.6	3.1	0.1	2.2	1.9
Wheat	1.0	3.0	0.02	2.6	—
Soybean	1.0	3.0	0.05	2.3	—
	1.0	6.0	0.05	3.6	—

such canopies, Eq. (5.4) should only be applied to the top layers (30 to 40 percent) of the canopy.

Example 5.3. What is the wind speed at a height of 1 m in a 2 m high corn crop if the wind speed 1 m above the top of the crop is 4.6 m/s?

Solution. From the information given, $z = 3$ m, $d = 1.4$ m, $z_m = 0.2$ m, and $u(3) = 4.6$ m/s. Using Eq. (5.1), gives $u^* = 0.88$ m/s. Using the result of Examples 2, $u(h) = 2.75u^* = 2.43$ m/s. Now, using Eq. (5.4) with $a = 2$ (for corn) gives:

$$u(1) = 2.43 \exp\left[2\left(\frac{1}{2} - 1\right)\right] = 0.9 \text{ m/s.}$$

References

Bussinger, J. A. (1975) Aerodynamics of vegetated surfaces. Heat and Mass Transfer in the Biosphere (D. A. de Vries and N. H. Afgan, eds.) New York; Wiley.

Cionco, R. M. (1972) A wind-profile index for canopy flow. Boundary-Layer Meteorol. 3:255-263.

Gifford, F. A. (1968) An outline of theories of diffusion in the lower layers of the atmosphere. Meteorology and Atomic Energy (D. H. Slade, ed.). USAEC Division of Technical Information Extension, Oak Ridge, Tenn.

Goudriaan, J. (1977) Crop Micrometeorology: A Simulation Study. Centre for Agricultural Publishing and Documentation, Wageningen, The Netherlands. 249 pp.

Hansen, F. V. (1993) Surface Roughness Lengths. ARL Technical Report, U. S. Army, White Sands Missile Range, NM 88002-5501.

Inoue, K. and Z. Uchijima (1979) Experimental study of microstructure of wind turbulence in rice and maize canopies. Bull. Nat. Inst. Agric. Sci. Tokyo, Japan, Ser. A 26:1-88.

Lowry, W. P. (1969) Weather and Life, New York: Academic Press.

Lumley, J. L. and H. A. Panofsky (1964) The Structure of Atmospheric Turbulence. New York: Wiley.

Maki, T. (1975) Interrelationships between zero-plane displacement, aerodynamic roughness length and plant canopy height. J. Agricultural Meteorology 31:61-70. (Japanese text, English summary and figure legends.)

Monteith, J.L. (1973) Principles of Environmental Physics. New York: American Elsevier.

Sauer, T. (1993) Sensible and Latent Heat Exchange at the Soil Surface Beneath a Maize Canopy. Ph.D. thesis, Department of Soil Science, Univ. of Wisconsin, Madison, WI.

Shaw, R. H. and A. R. Pereira (1982) Aerodynamic roughness of a plant canopy: a numerical experiment. Agric. Meteorol. 26:51-65.

Tatarski, V. I. (1961) Wave Propagation in a Turbulent Medium, New York: Dover.

Problems

5.1. If the wind measured at 2 m (height measured from the soil surface) over a wheat field is 5 m/s, what is the wind speed at the top of the canopy? Assume the wheat is 60 cm in height.

5.2. Using the following data find u^* and z_m. Assume $d = 0.08$ m. Use a graphical or curve-fitting method; not Eqs. (5.2) and (5.3).

Height (m)	Wind speed (m/s)
0.5	7.0
1.0	8.3
2.0	9.4
4.0	10.6

5.3. What is the wind speed 10 cm above the ground in problem 5.1?

Heat and Mass Transport

6

Life depends on heat and mass transfer between organisms and their surroundings. Such processes as carbon dioxide exchange between leaves and the atmosphere, oxygen uptake by micro-organisms, oxygen and carbon dioxide exchange in the lungs of animals, or convective heat loss from the surfaces of animal coats are fundamental to the existence of life. A thorough understanding of these exchange processes is therefore a necessary part of the study of biophysical ecology.

Now that we have the ability to describe the state of the organism environment (temperature, gas concentrations, wind, etc.) and the temperature or concentration at the organism surface, we are ready to show how to compute fluxes of heat and mass between organisms and their surroundings. In this section we briefly review the transport laws, and then show how to integrate the laws of heat and mass transfer to obtain forms of them that are easily used in environmental biophysics. We then show how to apply these laws to compute fluxes. In Ch. 7 we use principles from engineering and micrometeorology to compute conductances for organisms and their surroundings, and then, in Chs. 8 and 9 we briefly present the theory for heat and water flow in soil.

The rate of transport of mass or energy is usually expressed as the product of a proportionality factor and a driving force. The most familiar transport laws are the following:

Newton's law of viscosity for momentum transport:

$$\tau = \mu \frac{du}{dz}.$$ (6.1)

Fick's law for diffusive transport of material:

$$F'_j = -D_j \frac{d\rho_j}{dz}.$$ (6.2)

Fourier's law for heat transport:

$$H = -k \frac{dT}{dz}.$$ (6.3)

Darcy's law for fluid (water) flow in a porous medium (soil):

$$J_w = -K(\psi) \frac{d\psi}{dz}. \tag{6.4}$$

In Eqs. (6.1) through (6.4) τ is the shear stress (N/m^2) between layers of a moving fluid with dynamic viscosity μ and velocity gradient du/dz; F'_j is the flux density (kg m^{-2}s^{-1}) of a diffusing substance with molecular diffusivity D_j (m^2/s) and concentration or density (kg/m^3) gradient of $d\rho_j/dz$; H is the heat flux density (W/m^2) in a substance with thermal conductivity k $(Wm^{-1}K^{-1})$ and temperature gradient dT/dz (C/m); and J_w is the water flux density (kg m^{-2} s^{-1}) in soil with hydraulic conductivity $K(\psi)$ (kg s m^{-3}) and water potential gradient $d\psi/dz$ (J kg^{-1} m^{-1} or m/s^2). We use Eq. (6.2) mainly for describing diffusion of gases in air. The subscript j represents the different substances that diffuse through air. Here we are concerned mainly with water vapor, CO_2, and oxygen for which we use the subscripts v, c and o. The negative signs in Eqs. (6.2) through (6.4) indicate that the flux is in the positive direction when the gradient is negative.

The strong dependence of hydraulic conductivity on water potential in unsaturated soil is indicated by $K(\psi)$ in Eq. (6.4). The other coefficients are almost constant. Equations (6.1) through (6.3) therefore express a nearly linear relationship between a flux density and a driving force (a "concentration" gradient).

6.1 Molar Fluxes

In order for us to use the mass transport equation, it needs to be converted to the form for molar fluxes. Substituting Eq. (3.2) into 6.2 gives

$$F_j = \frac{F'_j}{M_j} = -\hat{\rho} D_j \frac{dC_j}{dz} \tag{6.5}$$

where F_j is in mol m^{-2} s^{-1}. This is the form we use throughout this book. Note, however, that mole fluxes are easily converted to mass fluxes through multiplication by the molecular mass of the diffusing gas.

There are several advantages to expressing the heat equation in the same form as Eq. (6.5). The mathematical manipulations are then the same for both transport processes, and the units (m^2/s) are the same for both diffusivities. The diffusivities are roughly the same size and they have similar temperature and pressure dependence (which can be derived from kinetic theory). For many conditions of interest in environmental biophysics, the diffusivities are constant multiples of each other so if one is known, the other is easily found. Equation (6.3) can be converted to a form similar to Eq. (6.5) by multiplying and dividing by c_p, where c_p is the molar specific heat of air (29.3 J mol^{-1} C^{-1}). The quantity k/c_p is the thermal diffusivity D_H, so Eq. (6.3) becomes

$$H = -\hat{\rho} c_p D_H \frac{dT}{dz}. \tag{6.6}$$

6.2 Integration of the Transport Equations

In addition to the transport laws previously mentioned, a fifth law is in common use that describes the flow of current in an electrical circuit. This is Ohm's law, which states that the current flowing in a conductor is directly proportional to the applied voltage and inversely proportional to the electrical resistance of the conductor. This law is different from the other laws in that it applies to a macroscopic system. The applied voltage is measured across an entire conductor, not over an infinitesimal increment as is indicated by the differentials in Eqs. (6.1) through (6.6). The resistance or conductance of the conductor is a function of size and shape as well as basic material properties.

In environmental biophysics, our problems are similar to the circuit problem. It is usually possible to specify concentrations at the organism surface and in the surroundings some distance from the organism, but it is usually impossible to measure gradients on a microscopic scale, as would be needed for the transport equations that we have presented so far. We therefore write the transport equations, by analogy with Ohm's law, in an integrated or macroscopic form similar to Eq. (1.1). The concentrations are specified at the organism surface and in the surroundings, and the transport resistance or conductance is defined as the concentration difference divided by the flux density. The mass and heat flux equations are the ones we most often use in this form. Expressing Eqs. (6.5) and (6.6) in this form gives:

$$F_j = g_j(C_{js} - C_{ja}) = \frac{C_{js} - C_{ja}}{r_j} \tag{6.7}$$

and

$$H = g_H c_p(T_s - T_a) = \frac{c_p(T_s - T_a)}{r_H} \tag{6.8}$$

where g is the conductance (mol m^{-2} s^{-1}) and r is resistance (m^2 s/mol). For the simple case of pure, linear diffusion, g is just $\hat{\rho}D_j/\Delta z$. The resistance is always just the reciprocal of the conductance. The integrated forms Eqs. (6.7) and (6.8) are useful for many cases besides the simple one. Chapter 7 deals in more detail with integration of the transport equations to determine resistance values from basic fluid properties and system geometry.

6.3 Resistances and Conductances

As we have shown, it is convenient to express exchange of heat and mass between organisms and their environment in terms of a concentration difference multiplied by a conductance or divided by a resistance. The use of resistance for calculating heat and mass exchange is convenient because a series of several resistances are often found between the surface of the organism and the environment, and therefore the familiar series resistor formulas from electronics can be used to calculate the total resistance.

Conductances, however, are better for suggesting significance since the flux is directly proportional to the conductance. Conductances, rather than resistances, should be used for statistical operations, since errors in conductance, rather than the resistance are normally distributed. A mean conductance is therefore a reliable indicator of expected behavior of a population of exchange surfaces, while a mean resistance may be meaningless. This is easy to see if you consider two surfaces; one has infinite resistance, or zero conductance and the other has a resistance of 5 m^2 s/mol, or a conductance of 200 mmol m^{-2} s^{-1}. The average resistance of the two is infinite, suggesting that the average water loss is zero; but the average conductance is 100 mmol m^{-2} s^{-1}, suggesting the correct average water loss.

We use units of m^2 s/mol for resistance, and mol m^{-2} s^{-1} for conductance. Units for resistance in the previous edition of this book and in much of the environmental biophysics literature are s/m. To convert from s/m to m^2 s/mol, divide the s/m resistance by the molar volume of air, 41.4 mol m^{-3} (at sea level and 20° C).

6.4 Resistors and Conductors in Series

Consider the flow of heat from the core of an animal to the surrounding air. The resistances to heat flow are shown in Fig. 6.1. Here T_a is air temperature, T_s is the temperature at the surface of the coat, T_o is the skin surface temperature, and T_b is the body core temperature. The resistances shown are those for the boundary layer of air, for the coat, and for the tissue. Following the rules for determining the total equivalent resistance of resistors in series, the resistance for heat transfer from the body core to the air is

$$r_H = r_{Ha} + r_{Hc} + r_{Ht}. \qquad (6.9)$$

The total conductance is the reciprocal of the total resistance, and is therefore the reciprocal of the sum of the reciprocals of the component conductances:

$$g_H = \frac{1}{\frac{1}{g_{Ha}} + \frac{1}{g_{Hc}} + \frac{1}{g_{Ht}}}. \qquad (6.10)$$

The heat flux density through all resistors is the same, so the ratio of the temperature drop across any resistor to the total temperature difference between the body and air is equal to the ratio of that resistance to the total

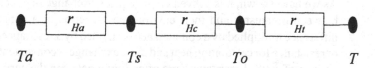

FIGURE 6.1. Resistors in series: resistances and temperatures from the core of an animal to the surrounding air.

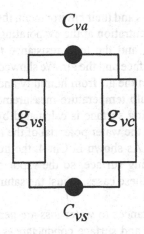

FIGURE 6.2. Conductors in parallel: vapor diffusion through the surface of a leaf.

resistance. If the total resistance were 10 m² s/mol, the coat resistance were 6 m² s/mol, $T_b = 37°$ C and $T_a = 0°$ C, the temperature change from the skin surface to the outer edge of the coat would be $(6/10) \times (37 - 0) = 22.2°$ C.

6.5 Resistors in Parallel

Now consider the loss of water from a leaf surface. Water can be lost both through the stomata and directly through the cuticle of the leaf. These represent parallel pathways for vapor loss, as shown in Fig. 6.2. Here C_{vs} and C_{va} represent water vapor concentration just inside and outside the leaf epidermis. The conductances are for cuticle and stomata. From the rules for parallel resistors in electronics, it is known that the combined resistance of resistors in a parallel circuit is the reciprocal of the sum of the reciprocals of the component resistors. The conductances simply add. The total conductance for the leaf is therefore

$$g_v = g_{vc} + g_{vs}. \tag{6.11}$$

The total, or equivalent resistance is:

$$r_v = \frac{1}{\frac{1}{r_{vc}} + \frac{1}{r_{vs}}}. \tag{6.12}$$

The vapor concentration difference across the two resistances is the same, and the vapor flux through each is proportional to the conductance of each resistor.

6.6 Calculation of Fluxes

In order to use Eq. (6.7) to calculate the rate of vapor exchange or latent heat exchange between plant canopies and the atmosphere, or between living organisms and their environment, the following needs to be known: the vapor concentration at the evaporating surface, the vapor concentration of the air, and the total resistance to vapor transfer between the evaporating surface and the air. We showed earlier how to calculate vapor concentration in the air from humidity and air temperature, or from dew point or wet bulb temperature measurements. The vapor concentration at the evaporating surface is calculated by knowing the temperature of the surface and the water potential of the liquid phase from which water is evaporating. As shown in Ch. 4, the humidity is often very near 1.0 at the evaporating surface, so the vapor concentration at the evaporating surface, in these cases, is just the saturation concentration at surface temperature.

The conductances to water loss are generally series combinations of boundary layer and surface conductances. We consider boundary layer conductance in detail in Ch. 7. To give some indication of the sizes of conductances in nature, a 1 cm thickness of still air has a conductance around 100 mmol m^{-2} s^{-1}. The boundary layer conductances of leaves and of crops typically range from 500 to 1000 mmol m^{-2} s^{-1}.

Equation (6.8) is used similarly. The temperature of the environment and organism needs to be known as well as values for all resistances or conductances between the organism and the environment. The total resistance may be a series combination of several component resistances, as it is with vapor resistances. The magnitudes of these resistances, at least in the boundary layer and in a layer of still air, are similar to those for vapor. We now go through several example calculations to show how Eqs. (6.7) and (6.8) are used.

Example 6.1. Find the rate of water loss from a crop. Assume the canopy temperature is 30° C, the air vapor pressure is 1.0 kPa, canopy conductance is 1 mol m^{-2} s^{-1}, and boundary layer conductance is 0.5 mol m^{-2} s^{-1}.

Solution. The humidity at the evaporating surfaces inside the leaves is essentially 1, (Eq. (4.13), with $\psi = -1000$ J/kg), so $C_{vs} = h_{rs}e_s(T_s)/p_a = 1 \times 4.24 kPa/101 kPa = 0.042$ mol/mol. The vapor concentration in the air is $C_{va} = 1.0 kPa/101 kPa = 0.0099$. The total conductance for vapor exchange is the series combination of canopy and boundary layer conductance:

$$g_v = \frac{1}{\frac{1}{1 \text{ mol m}^{-2} \text{s}^{-1}} + \frac{1}{0.5 \text{ mol m}^{-2} \text{s}^{-1}}} = 0.33 \text{ mol m}^{-2} \text{ s}^{-1}.$$

The evaporative loss is therefore

$$E = 0.33 \text{ mol m}^{-2} \text{ s}^{-1}(0.042 - 0.009)$$

$$= 0.0107 \text{mol m}^{-2} \text{ s}^{-1} = 10.7 \text{m mol m}^{-2} \text{ s}^{-1}.$$

The mass flux density is

$$0.0107 \frac{mol}{m^2 s} \times \frac{0.018 kg}{mol} = 1.92 \times 10^{-4} \frac{kg}{m^2 s}.$$

To get some feeling for the magnitude of this number, if evaporation continued at this rate for an hour, $3600 \, s \times 0.000192 \, kg \, m^{-2} \, s^{-1} = 0.7 \, kg/m^2$ would be evaporated. One kg/m^2 is 1 mm depth of water over one square meter. Therefore, 0.7 mm of water would have evaporated in an hour. The heat required to evaporate this amount of water is

$$\lambda E = 0.0107 \frac{mol}{m^2 s} \times 44000 \frac{J}{mol} = 471 \frac{W}{m^2}$$

(recall from Ch. 1 that a joule per second is equal to a watt).

Evaporation from a wet soil surface is similar to evaporation from a crop. When the soil surface is wet, water potential is near zero, and the vapor pressure at the surface is near saturation. As the soil dries, the resistances change drastically. The wet front, or point in the soil where $h_r = 1$ retreats into the soil, and the total resistance (the sum of the diffusion resistance through the soil and through the boundary layer of air above the soil) increases. A 1 cm thick layer of dry soil has a diffusive conductance for vapor of about $0.03 \, mol \, m^{-2} \, s^{-1}$.

Example 6.2. What is the rate of evaporation from a moist soil which is covered with a 5 cm thick dry soil layer? Assume the same surface temperature and air vapor pressure conditions as in the previous example.

Solution. The soil conductance is $0.03 \, mol \, m^{-2} \, s^{-1}/5 = 0.006 \, mol \, m^{-2} \, s^{-1}$. Assume the temperature of the wet soil below the dry layer is 30° C, similar to the surface in the previous example. The vapor flux calculation is like the previous example, but with soil conductance in series with boundary layer conductance. The overall vapor conductance is 0.0059 $mol \, m^{-2} \, s^{-1}$, and the evaporation rate is

$$E = 0.0059 \, mol \, m^{-2} s^{-1} (0.042 - 0.0099) = 0.2 \, mmol \, m^{-2} s^{-1}.$$

The water loss from the crop (or a wet soil) is therefore 50 times as great as that for the dry soil surface. This gives some indication of the effectiveness of a dry soil layer in slowing evaporation.

Example 6.3. Find the vapor conductance (skin plus boundary layer) of a potato if, when left for 12 hours on a laboratory bench, it lost 3 g of water. The tuber and laboratory are at 22° C, and the laboratory humidity is 0.53. The surface area of the potato is $310 \, cm^2$.

Solution. Rearranging Eq. (6.7) to find conductance gives

$$g_v = \frac{E}{C_{vs} - C_{va}}.$$

The evaporation rate is

$$E = \frac{\text{mass loss}}{\text{area} \times \text{time}} = \frac{\frac{3g}{12hr} \times \frac{1hr}{3600s} \times \frac{1mol}{18g}}{310 cm^2 \times \frac{1m^2}{100^2 cm^2}} = 0.000124 \, \frac{mol}{m^2 s}.$$

From Table A.3, the saturation vapor pressure at $22°$ C is 2.64 kPa. From Ch. 3, the mole fraction difference is

$$C_{vs} - C_{va} = \frac{e_s(T_a)(1 - h_r)}{p_a} = \frac{2.64k \, Pa(1 - 0.53)}{101k \, Pa} = 0.012 \, \frac{mol}{mol}.$$

The conductance is

$$g_v = \frac{0.00124 \, \frac{mol}{m^2 s}}{0.012 \, \frac{mol}{mol}} = 10.3 \, \frac{mmol}{m^2 s}.$$

Example 6.4. Warm moist winds can quickly melt substantial depths of snow. Heat transfer to the snow is latent as well as sensible. Compare the latent and sensible heat fluxes to a snow drift from saturated air at $5°$ C, if the boundary layer conductance is 1 mol m^{-2} s^{-1}.

Solution. From Table A.3, the saturation vapor pressure at $5°$ C is 0.87 kPa, and at $0°$ C (the surface temperature of the melting snow) it is 0.61 kPa. The sensible heat flux density is

$$H = 29.3 \, \frac{J}{mol \, C} \times 1 \, \frac{mol}{m^2 s} (0° \, C - 5° \, C) = -147 \, \frac{W}{m^2}$$

The latent heat flux density is the latent heat of vaporization multiplied by Eq. (6.7):

$$\lambda E = 44000 \, \frac{J}{mol} \times 1 \, \frac{mol}{m^2 s} \left(\frac{0.61k \, Pa - 0.87k \, Pa}{101k \, Pa} \right) = -114 \, \frac{W}{m^2}.$$

The negative signs indicate that the flux is toward the surface. The total heat flux to the surface is 261 W/m^2. The interesting thing about this computation is that the latent heat flux is almost half of the total.

Example 6.5. A person has a sleeping bag which has a thermal conductance of 0.05 mol m^{-2} s^{-1}. The tissue conductance of the person, while

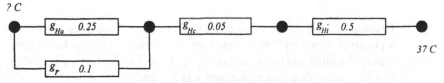

FIGURE FOR EXAMPLE 6.5.

sleeping, is around $0.5 \, mol \, m^{-2} \, s^{-1}$. Heat exchange with the surroundings is by radiation and convection in parallel. The convective conductance is $0.25 \, mol \, m^{-2} \, s^{-1}$ and the radiative conductance is $0.1 \, mol \, m^{-2} \, s^{-1}$. What is the coldest temperature that the sleeping bag can be used at if the person maintains a metabolic rate of $50 \, W/m^2$ and a body temperature of $37° \, C$?

Solution. The conductor network for the sleeping bag and surroundings is shown in the diagram. Equations (6.10) and (6.11) can be used to calculate the total thermal conductance. The overall conductance is:

$$g_H = \frac{1}{\frac{1}{0.25+0.1} + \frac{1}{0.05} + \frac{1}{0.5}} = 0.04 \, mol \, m^{-2} \, s^{-1}.$$

Now rearrange Eq. (6.8) to find the required temperature:

$$T_a = T_b - \frac{H}{g_H c_p} = 37° \, C - \frac{50 \frac{W}{m^2}}{0.04 \frac{mol}{m^2 s} \times 29.3 \frac{J}{mol \, C}} = -5.5° \, C$$

If the air temperature drops below $-5.5° \, C$, either the person's body temperature will drop, or the metabolic rate must increase.

Problems

6.1. A leaf is at $27° \, C$ and is in air at $30° \, C$ with 0.2 relative humidity. The stomatal conductance is $250 \, mmol \, m^{-2} \, s^{-1}$ and the boundary layer conductance is $900 \, mmol \, m^{-2} \, s^{-1}$. What is the rate of water loss?

6.2. What is the heat flux density in problem 6.1? Assume the boundary layer conductance for heat is the same as for vapor.

6.3. Your body temperature is $37° \, C$. If your average heat loss while sitting in a $22° \, C$ room is $80 \, W/m^2$, what is the total (boundary layer, clothing, and tissue) conductance? If your average tissue conductance is $1 \, mol \, m^{-2} \, s^{-1}$, what is your average skin temperature?

Conductances for Heat and Mass Transfer

7

This chapter continues the discussion of transport, and focuses on methods for computing the conductances and resistances needed for the calculations in Ch. 6. We first discuss conductances on the smallest spatial scale; molecular diffusion. It is by this process that heat and mass are transported in still air or water, such as in parts of the lungs of animals, in soils, in the substomatal cavities of leaves, and in animal coats. The equations for turbulent transport of heat and mass on larger scales in the atmosphere are similar to those for molecular diffusion, so those equations are discussed following the molecular diffusion equations. After diffusion processes are discussed, we consider an intermediate scale; namely, convective heat and mass transfer theory as it applies to fluids moving over plates, cylinders, and spheres (simulating leaves, stems, fruits, and animals).

7.1 Conductances for Molecular Diffusion

To determine a conductance for molecular transport, we return to Fick's law for steady diffusion of some component j (Eq. (6.5)):

$$F_j(z) = -\hat{\rho} D_j \frac{dC_j}{dz} \tag{7.1}$$

where D_j is the diffusivity and dC_j/dz is the concentration gradient. Assume that all of the material that diffuses across an imaginary boundary at z originated at a surface with area $A(z_s)$ where the flux density is $F(z_s)$, which is assumed to be constant, then:

$$F_j(z_s) = \frac{A(z)}{A(z_s)} F_j(z) = -\frac{A(z)}{A(z_s)} \hat{\rho} D_j \frac{dC_j}{dz} \tag{7.2}$$

where $A(z)$ is the area of the surface at z. Integration of Eq. (7.2), combining the result with Eq. (6.7), and solving for conductance gives:

$$g_j = \frac{\hat{\rho} D_j}{A(z_s) \int_{z_s}^{z_a} \frac{dz}{A(z)}} \tag{7.3}$$

where z_a is at the outer edge of the layer for which the diffusive conductance is being computed. Equation (7.3) is easily solved for several simple shapes. For planar diffusion, such as diffusion through long, narrow tubes or from large, plane surfaces $A(z) = A(z_s)$. The integration is then trivial, giving

$$g_j = \frac{\hat{\rho} D_j}{\Delta z} \tag{7.4}$$

with Δz being the distance from the source ($z_s = 0$) to the point at which C_j is measured. The decrease in concentration is linear with distance from the surface.

For diffusion from a spherical surface, $A(z) = 4\pi z^2$ (z is the radial distance from the center of the sphere). Using Eq. (7.3) gives the conductance at a distance z_a from the center of the sphere. It is:

$$g_j = \frac{\hat{\rho} D_j}{z_s \left(1 - \frac{z_s}{z_a}\right)} \tag{7.5}$$

where the radius of the spherical surface is z_s. In the limit, as z_a becomes very large, the ratio of the radii in the denominator approaches zero and the conductance approaches the value for planar diffusion through a distance equal to the radius of the sphere.

For a cylindrical surface with unit length $A(z) = 2\pi z$ (z is the radius of the cylinder). Integration of Eq. (7.3) gives:

$$g_j = \frac{\hat{\rho} D_j}{z_s \ln \frac{z_a}{z_s}} \tag{7.6}$$

where z_s is the radius of the cylindrical exchange surface and z_a is the distance from the cylinder axis to the point of concentration measurement. The logarithm term does not approach any limit as z_a increases, so there is no lower bound to the conductance of a cylinder as there is with a sphere. The rate of decrease with distance from the surface does become small, however, at large distances. It is interesting to note the similarity in form among the three conductance equations. Each has a density times a diffusivity divided by a length. In the case of the sphere and the cylinder the length is the radius of the object multiplied by a factor. The factor ranges from 0 to 1 for the sphere. For the cylinder, the theoretical upper limit is infinity, but the practical upper limit is 5 or 6. Equation (7.3) could, of course, be integrated for other shapes, but these three cover most situations of interest to us in this book.

7.2 Molecular Diffusivities

Before using Eqs. (7.4) through (7.6), values for D_j are needed. These depend on the properties of the diffusing substance and the medium in which diffusion occurs. Molecular diffusion coefficients for heat, water vapor, oxygen, and CO_2 in air are given in Table A.1. Available data for diffusion in water are given in Table A.2. The diffusivities in air at 20° C are also shown in Table 7.1.

TABLE 7.1. Viscosity and diffusivities of heat, water vapor, carbon dioxide, and oxygen for air at 20° C and 101 kPa pressure.

	Molecular weight (g/mol)	Diffusivity (m²/s)
viscosity	—	1.51×10^{-5}
heat	—	2.14×10^{-5}
water vapor	18.02	2.40×10^{-5}
carbon dioxide	44.01	1.57×10^{-5}
oxygen	32.00	2.00×10^{-5}

The molecular processes which cause diffusion result in similar values for all of the diffusivities. For mass transport, Graham's law states that the ratio of the diffusivities is equal to the inverse of the square root of the ratio of the molecular weights. Comparing diffusivities of water vapor, oxygen, and carbon dioxide, it can be seen that their ratios approximate the predictions from Graham's law. Carbon dioxide has the largest molecular weight and has the smallest diffusivity.

Diffusivity changes with temperature and atmospheric pressure. Fuller et al. (1966) suggest the following

$$D_j(T, p_a) = D_j(293.16 \text{ K}, 101.3 \text{ kPa}) \frac{101.3}{p_a} \left(\frac{T}{293.16} \right)^{1.75} \quad (7.7)$$

where $D_j(293.16$ K, 101.3 kPa$)$ represents the appropriate diffusivity from Table 7.1 and p_a is the atmospheric pressure from Eq. (3.7). Substituting this into Eqs. (7.4) through (7.6), and using Eq. (3.3) for the density, shows one of the big advantages of using mole fractions in the transport equations. The pressure terms divide out, so that there is no pressure dependence of the conductance, and much of the temperature dependence divides out so that there is only 0.25%/C left, which can often be neglected. (Note: The 0.25%/C comes from a dependence on $T^{0.75}$ so that $(301^{0.75} - 300^{0.75})/(300.5^{0.75}) = 0.0025$).

Example 7.1. The finger of a wool glove has a diameter of 3 cm. The diameter of a person's finger inside the glove is 2 cm. If the wool acts like a layer of still air around the finger, what is the conductance of the glove finger at 20° C and 100 kPa?

Solution. The finger approximates a cylinder with $z_s = 0.01$ m and $z_a = 0.015$ m. Using Eq. (7.6), the conductance is:

$$g_H = \frac{41.0 \frac{\text{mol}}{\text{m}^3} \times 2.14 \times 10^{-5} \frac{\text{m}^2}{\text{s}}}{0.01 \text{ m } \ln \frac{0.015 \text{ m}}{0.01 \text{ m}}} = 0.219 \frac{\text{mol}}{\text{m}^2\text{s}}.$$

Using Eq. (7.4) the conductance of a similar thickness of still air in a plane can be computed. The conductance is 0.177 mol m^{-2} s^{-1}, about 80 percent of the cylindrical value, so curving the insulation around the finger results in 20 percent more heat loss than if the material were flat. This is one reason why a mitten keeps fingers warmer than a glove. The actual conductance of a fiber layer like this can be more than twice the conductance of a similar thickness of still air because of radiative and convective transport within the material, but the conduction through the air provides a good starting point for the calculation of overall conductance. We return to this subject later after developing the tools to analyze these other modes of heat transport.

7.3 Diffusive Conductance of the Integument

The waterproof coating that covers most forms of terrestrial life plays a key role in maintaining a favorable water balance. The most effective of these coatings are made up of lipids or waxes, but layers of hair and other dry materials also impede evaporation. We consider three situations involving gas diffusion through the integument. The first is diffusion through a layer of still air, such as an animal coat. The second is diffusion through a cuticle made up of lipid layers. The third is diffusion through pores in a cuticle.

The first case is one that clearly involves diffusion of gases in air, so the equations just derived apply. The presence of the hair tends to keep the air still and impede convection, but the fraction of volume taken up by the hair has little effect on the area available for diffusion. In the second case, the water is not diffusing in air, but through the lipid layers of the cuticle. The proper diffusivity to use is therefore not the one for air, but the one for the membrane through which diffusion is occurring. The driving forces, however, are the same as those for diffusion in air, and conductances are obtained simply by using measured rates of water loss and vapor concentration differences. We are not able to derive equations to compute values for these conductances, but their values tend to be conservative (i.e., do not change with ambient conditions) so observed values are useful for calculations. Table 7.2 gives a sample of values for arthropod, animal, and plant surfaces.

An example of the third case is the transport of gases through stomata in leaves. The conductance of a single stomatal pore is given by Eq. (7.4), where Δz is the pore depth. To account for nonplanar diffusion just outside the stomatal pore an end correction is applied. The overall conductance of the perforated surface is given by (deMichael and Sharp, 1973):

$$g_{js} = \frac{n\hat{\rho}D_j}{\left(\frac{\Delta z}{A} + \frac{\pi}{2L_o}\right)} \tag{7.8}$$

where A is the area of a single pore, n is the number of pores per square meter, and L_o is the pore perimeter. Equation (7.8) is valid for water vapor, CO_2, or oxygen when the appropriate diffusion coefficient is used.

TABLE 7.2.   Integument vapor conductances (from Monteith and Campbell, 1980; Monteith and Unsworth, 1990; and Patten et al., 1988).

Arthropods	mmol m^{-2} s^{-1}	Birds	mmol m^{-2} s^{-1}	
Lithobius sp.	32.	*Melopsittacus indulatus*	4.9	
Porcellio scaber	13.	*Excalifactoria chinensis*	2.1	
Hemilepistus reaumuri	2.8	*Eggs, several species*	0.55	
Glossinia palpalis	1.4			
Ornithodorus maubata	0.48	**Vegetables & fruits**		
Androctonus australis	0.096	*potato tuber*	0.77	
		apple, Red Delicious	1.2	
Mammals		*apple, Golden Delicious*	2.4	
Homo sapiens	5.4	*Tomato*	5.5	
Acomys sp.	2.8	*Orange*	5.8	
		Radish	275	
Reptiles				
Caiman sp.	7.5	**Plant leaves**	open	closed
Terrapene sp.	1.3	*Beta vulgaris*	260	10
Gopherus sp.	0.35	*Gossypium hirsutum*	375	13
		Betulua verrucosa	360	5.9
		Pinus monticola	330	17
		maize	330	30
		soybean	450	40
		Quercus robur	41	2.1

Interactions between diffusing species and convection corrections are important in some studies (Jarman, 1974) but are not discussed in detail here.

In practice, Eq. (7.8) is seldom used to find stomatal conductance because it is harder to determine pore diameters, lengths, and numbers than it is to directly measure the stomatal conductance. Table 7.2 gives examples of stomatal conductances of leaves, for both open and closed stomata. When stomata are tightly closed, the diffusive conductance is mainly the conductance of the cuticle.

Example 7.2. Leaves of some species have a thick pubescent layer over the entire surface of the leaf. If this layer creates a 1 mm thickness of still air over the surface of the leaf, how does the conductance of that layer compare with the stomatal conductance of a typical leaf?

Solution. Using Eq. (7.4), the conductance of a 1 mm thick layer of still air is

$$g_{vc} = \frac{41.4 \, \frac{\text{mol}}{\text{m}^3} \times 2.4 \times 10^{-5} \, \frac{\text{m}^2}{\text{s}}}{0.001 \, \text{m}} = 0.99 \, \frac{\text{mol}}{\text{m}^2 s}.$$

The resistance is $r_{vc} = 1/g_{vc} = 1/0.99 = 1.01$ m^2 s/mol. Stomatal conductances for several of the species listed in Table 7.2 are around 300 mmol m^{-2} s^{-1} when open and 10 mmol m^{-2} s^{-1} when closed. The corresponding resistances are 3.3 and 100 m^2 s/mol. The resistance of the stomata plus pubescence is therefore 4.3 m^2 s/mol for open stomata and 101 m^2 s/mol for closed stomata. The pubescence could therefore decrease transpiration by about 25 percent when stomata are open, but would have a negligible effect when stomata are closed.

Example 7.3. Consider stomatal density on upper and lower sides of corn, oat, and bean leaves and calculate the leaf stomatal conductance using number density (n), pore length (l), and pore width (w).

Species	stomata per mm² adaxial side	stomata per mm² abaxial side	length (μm)	width (μm)
bean	40	280	13	7
corn	52	68	19	5
oat	25	23	38	8

Solution. If the stomata are considered ellipses where

$$A = \pi(l/2)(w/2) \qquad L_o = 2\pi\sqrt{[(l/2)^2 + (w/2)^2]/2}$$

and the pore depth is assumed to be $\Delta z = 10\mu$m, then for bean leaves:

$$A = 71.5 \times 10^{-12}\,m^2$$
$$L_o = 32.8 \times 10^{-6}\,m$$

$$g_{vs}(\text{adaxial}) = \frac{40 \times 10^{-6}\,m^{-2} \times 2.4 \times 10^{-5}\,m^2\,s^{-1} \times 41.4\,mol\,m^{-3}}{\frac{10 \times 10^{-6}\,m}{71.5 \times 10^{-12}\,m^2} + \frac{3.14}{2 \times 32.8 \times 10^{-6}\,m}}$$

$$= 0.212\,\frac{mol}{m^2 s}$$

$$g_{vs}(\text{abaxial}) = 1.47\,mol\,m^{-2}\,s^{-1}$$

$$g_{vs}(\text{total}) = 0.212 + 1.47 = 1.68\,mol\,m^{-2}\,s^{-1}.$$

For oat leaves:

$$g_{vs}(\text{total}) = 0.41 + 0.38 = 0.79\,mol\,m^{-2}\,s^{-1}.$$

The calculation for corn is left as an exercise for the reader. These conductances are considerably larger than values shown in Table 7.2. Obviously, direct comparisons are not warranted here because we do not know the stomatal density and sizes for the leaves in the table. It is possible, however, that stomata interact in some way not accounted for in Eq. (7.8). Alternatively, the disagreement between this calculation and measurements may arise from the presence of additional resistances to diffusion, including the substomatal cavity, or uncertainties in the measurements of pore size and density.

7.4 Turbulent Transport

We return now to the subject of Ch. 5 and consider the transport of heat and mass in the atmosphere by turbulence. The fluctuations, or eddies, in the atmosphere are, in a sense, like molecules in a gas. They bounce about with random motion, but are carried along with the wind. It is these fluctuations that transport heat, water, momentum, etc. in the atmosphere. If a packet of air at one level, with a given temperature and momentum, jumps to a different level in the atmosphere, the old heat and momentum are carried to the new level. This is analogous to the diffusion process in a gas, except that diffusion involves jumps of single molecules. Because heat, momentum, and mass are transported by the jumps between layers of these packets of air, the flux can be measured by averaging the product of fluctuations of temperature, horizontal wind, or mass, and vertical wind. This method of measuring fluxes is called eddy correlation or eddy covariance. The equations for determining the fluxes are:

$$\tau = -\rho \overline{u'w'} \tag{7.9}$$

$$H = \hat{\rho} c_p \overline{w'T'} \tag{7.10}$$

$$E = \hat{\rho} \overline{w'C_v'} \tag{7.11}$$

where τ is the momentum flux to the surface (or drag of the wind on the surface) often referred to as the shear stress, H is the heat flux density, E is the flux density of water vapor, and C_v' is the mole fraction given by e'/p_a. The primes indicate fluctuations about the mean, and overbars indicate averages taken over 15 to 30 minutes. A simple understanding of these equations can be obtained by considering the meaning of "fluctuations about the mean." In Eq. (7.9), if an eddy fluctuation is downward ($w' < 0$), then the horizontal wind fluctuation associated with this downward eddy will tend to be greater than the mean wind ($u' > 0$) because the horizontal wind speed tends to be larger at heigher heights (Fig. 5.3). Thus, downward moving eddies tend to carry higher horizontal wind speeds with them and upward moving eddies tend to carry lower horizontal wind speed upward into the faster moving stream. This means that the product $u'w'$, which is the covariance between u and w, is negative and we put a negative sign in Eq. (7.9) because by arbitrary convention we want to define a momentum flux toward the surface (in the negative z direction) as positive. Of course the correlation between u' and w' is not perfect. In general, the correlation between u' and w' ($u'w'/(\sigma_u \sigma_w)^{1/2}$) typically varies from about 0.1 to 0.4. The same kind of interpretation of vertical velocity fluctuations and temperature or gas concentration fluctuations is possible. Instruments must have a very fast response to make these measurements, and measurements must be sampled at least five to ten times per second to properly sample the eddies that are responsible for transport. If these requirements can be met, eddy correlation is a very attractive method for direct measurement of transport in the atmosphere.

In each of these flux equations, the transport is accomplished by fluctuations in the vertical wind component. The ability of the atmosphere to

transport heat or mass depends directly on the size of the vertical fluctuations. Because of the proximity of the surface, the magnitude of w'
is limited near the surface. Farther from the surface, transport is more
efficient because the eddies are larger. Transport is therefore expected to
increase with height. Increased intensity of turbulence will also increase
vertical transport. Since turbulence is generated by mechanical action of
wind moving over a rough surface, transport should increase as wind
speed and surface roughness increase. It is also known that turbulence is
generated by buoyancy, so when there is strong heating at the surface,
turbulent transport should increase. Strong cooling at the surface should
result in reduced transport.

Both the turbulence and the heat or mass being carried by it are generated or absorbed at the surface. The sources and sinks are patchy, so
near the surface the concentrations are not well related to the transport.
The mixture becomes more homogeneous with distance from the surface, and fluxes can then be predicted from concentration gradients. The
equations used for this are similar to those used for molecular transport,
but, of course, the mechanism for transport is quite different. We simply
define transport coefficients for turbulent diffusion that replace the molecular viscosity and diffusivities in Eqs. (6.1) through (6.3). The resulting
equations are the basis for what is known as K-theory. The equations are

$$\tau = K_M \rho \frac{du}{dz} \tag{7.12}$$

$$H = -K_H \hat{\rho} c_p \frac{dT}{dz} \tag{7.13}$$

$$E = -K_v \hat{\rho} \frac{dC_{va}}{dz} \tag{7.14}$$

where K_M is the eddy viscosity, K_H is the eddy thermal diffusivity, and
K_v is the eddy vapor diffusivity. These are steady-state flux equations
for the surface boundary layer of the atmosphere. The flux of heat or
mass is intuitive, but the flux of momentum is somewhat less intuitive
for most people. From introductory physics, momentum is mass times
velocity. With a solid object of a given mass that is moving at some
velocity, the momentum is obvious. With a fluid these quantities are less
obvious because the amount of mass depends on what volume is being
considered; further, a given mass of fluid can have a range of velocities.
Therefore with fluids the mass per unit volume is used instead of just
mass so what is referred to as momentum is actually momentum per unit
volume or concentration of momentum (ρu). The flux of any quantity
can be written as a product of the concentration of that quantity times
the appropriate velocity. In the surface layer, the appropriate velocity is
referred to as the friction velocity, which is represented by the symbol
$u^*(m/s)$, and is defined as $u^* = (\tau/\rho)^{1/2}$. Effects of wind speed, surface
roughness, and surface heating are all included in the shear stress term
(τ).

Equations (7.12) through (7.14) are not very useful in this form because there is no way of knowing what values to give the coefficients. From the discussion on turbulent mixing in the surface boundary layer, it is known that K will increase with height above the surface, wind speed, surface roughness, and heating at the surface. In the surface boundary layer, at steady state, the flux densities, τ, H, and E are assumed to be independent of height. Increases in the K coefficients with z will therefore be balanced by corresponding decreases in the gradients.

It can be assumed that K has some value, characterized by surface properties, at the exchange surface (where $z = d + z_M$ or $z = d + z_H$, etc.) and increases linearly with u^* and z. Based on these assumptions, the form of the Ks must be:

$$K_M = 0.4u^*(z - d)/\phi_M$$
$$K_H = 0.4u^*(z - d)/\phi_H \qquad (7.15)$$
$$K_v = 0.4u^*(z - d)/\phi_v.$$

As in Ch. 5, 0.4 is von Karman's constant. The ϕs are dimensionless influence factors which equal one for pure mechanical turbulence (no surface heating or cooling). These equations make the meaning of the roughness lengths more apparent. When $z = d + z_M$, K_M, is equal to $0.4u^* z_M$. The roughness length therefore just represents a characteristic length which makes the eddy viscosity equal to the value it has at the exchange surface.

If Eqs. (7.15) are substituted into Eqs. (7.12) through (7.14), and the resulting equations integrated from the height of the exchange surface $d + z_M$ to some height z the resulting equations describe the profiles of wind, temperature, and vapor concentration with negligible surface heating ($\varphi_M = \varphi_H = \varphi_v = 1$):

$$u(z) = \frac{u^*}{0.4} \ln \frac{z - d}{z_M} \qquad (7.16)$$

$$T(z) = T(d + z_h) - \frac{H}{0.4\hat{\rho}c_p u^*} \ln \frac{z - d}{z_H} \qquad (7.17)$$

$$C_{va} = C_{va}(d + z_v) - \frac{E}{0.4\hat{\rho}u^*} \ln \frac{z - d}{z_v}. \qquad (7.18)$$

The wind and temperature profile equations have been seen before in Chs. 2 and 5. Equations similar to these could be derived for other substances being transported by atmospheric turbulence (such as ozone, SO_2, volatile chemicals, etc.). Equations (7.16) through (7.18) represent flux-profile relationships in the atmospheric surface layer above soil or vegetation. Typically this surface layer extends to distances of 10 to 100 m above the surface. Above this surface layer is another layer referred to as the planetary boundary layer, which has quite different properties.

In Ch. 5 we discuss ways to determine values for z_M, the momentum roughness length. The roughness lengths for heat, vapor, and other scalars are assumed to be equal to each other, and are sometimes assumed to equal

z_M. However, the process of momentum exchange within the canopy differs from the scalar exchange processes (Garratt and Hicks, 1973), and these differences can be modeled by adjusting the scalar roughness parameters. For our computations of aerodynamic conductance we assume that:

$$z_H = z_v = 0.2 z_M. \tag{7.19}$$

7.5 Fetch and Buoyancy

Now that we have equations describing turbulent transport, we need to look briefly at their limitations. We started by assuming that the wind was *at a steady state* with the surface (that there were no horizontal gradients). When wind passes from one type of surface to another it must travel some distance before a layer of air, solely influenced by the new surface, is built up. The height of influence increases with downwind distance. The length of uniform surface over which the wind has blown is termed fetch, and the wind can usually be assumed to be 90 percent or more equilibrated with the new surface to heights of $0.01 \times$ fetch. Thus, at a distance 1000 m downwind from the edge of a uniform field of grain, expect the wind profile equations to be valid to heights of around 10 m.

The effect of thermally produced turbulence on transport was alluded to earlier, but its quantitative description was not given. The equations derived to this point apply only for mechanically produced turbulence, so they are appropriate only for adiabatic conditions. Strong heating of the air near the surface of the earth causes overturning of the air layers, with resultant increases in turbulence and mixing. Conversely, strong cooling of these air layers suppresses mixing and turbulence. Thus convective production or suppression of turbulence is directly related to sensible heat flux (H) at the surface. When H is positive (surface warmer than the air) the atmosphere is said to be unstable, and mixing is enhanced. When H is negative, the atmosphere is said to be stable, and mixing is suppressed by thermal stratification. When surface heating or cooling occurs, corrections to Eqs. (7.16) through (7.18) are made and referred to as "diabatic corrections."

The main components of a (random) kinetic energy budget for a steady-state atmosphere can be written as (Lumley and Panofsky, 1964):

$$\frac{-u^{*3}}{0.4z} + \frac{gH}{\hat{\rho}c_p \mathbf{T}} = \varepsilon \tag{7.20}$$

where g is the gravitational acceleration. The first term represents mechanical production of turbulent kinetic energy, the second term is the convective production, and these two together equal the viscous dissipation of the energy, ε. The ratio of convective to mechanical production of turbulence can be used as a measure of atmospheric stability:

$$\zeta = -\frac{0.4gzH}{\hat{\rho}c_p \mathbf{T}u^{*3}}. \tag{7.21}$$

The diabatic flux and profile equations can now be written as functions of stability and the other parameters already discussed. Only the wind and temperature equations are given. Fluxes and profiles of water vapor, carbon dioxide, and other scalars are similar to those for temperature.

For the diabatic case, the ϕs (diabatic influence factors) in Eqs. (7.15) increase from unity with positive ζ (stable atmosphere) and decrease with negative ζ. Yasuda (1988) gives the following equations.

For unstable conditions:

$$\phi_M = \frac{1}{(1 - 16\zeta)^{1/4}}; \quad \phi_H = \phi_M^2. \tag{7.22}$$

For stable conditions:

$$\phi_M = \phi_H = 1 + \frac{6\zeta}{1 + \zeta}. \tag{7.23}$$

The flux equations are integrated using these corrections to obtain the corrected profile equations. The diabatic profile equations are:

$$u(z) = \frac{u^*}{0.4} \left[\ln \left(\frac{z - d}{z_M} \right) + \Psi_M \right] \tag{7.24}$$

$$T(z) = T(d + z_H) - \frac{H}{0.4 \hat{\rho} c_p u^*} \left[\ln \left(\frac{z - d}{z_H} \right) + \Psi_H \right] \tag{7.25}$$

where Ψ_M and Ψ_H are the profile diabatic correction factors. The diabatic correction factors are zero for neutral conditions, and can be derived from the integration for the stable case. For unstable flow the integration cannot be carried out analytically, so it is done numerically and an empirical function is fit to the result. The profile diabatic correction factors are as follows.

For unstable flow:

$$\Psi_H = -2 \ln \left[\frac{1 + (1 - 16\zeta)^{1/2}}{2} \right]; \quad \Psi_M = 0.6 \Psi_H. \tag{7.26}$$

For stable flow:

$$\Psi_M = \Psi_H = 6 \ln(1 + \zeta). \tag{7.27}$$

The diabatic corrections are shown in Fig. 7.1.

7.6 Conductance of the Atmospheric Surface Layer

An important result of the previous section is the derivation of the profile equations which allow interpolation and extrapolation of atmospheric variables. Another important result is the development of an equation for computing the conductance of the atmospheric surface layer. Again, only the equation for heat is given, since the equations for all other scalars are

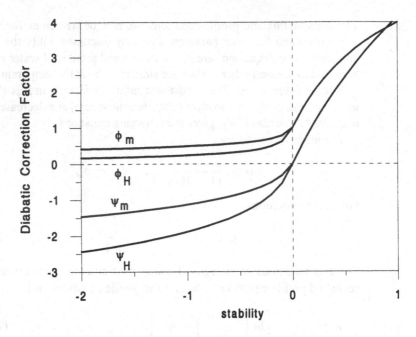

FIGURE 7.1. Diabatic influence and profile correction factors for heat and momentum as a function of stability (ζ).

the same. Rearranging Eq. (7.25) and substituting Eq. (7.24) for u^* gives:

$$H = \frac{0.4^2 \hat{\rho} c_p u(z)[T(d + z_H) - T(z)]}{\left[\ln\left(\frac{z-d}{z_M}\right) + \Psi_M\right]\left[\ln\left(\frac{z-d}{z_H}\right) + \Psi_H\right]}.$$

The conductance between the "canopy surface" (at height $d + z_H$) and a height z above the canopy is therefore

$$g_{Ha} = \frac{0.4^2 \hat{\rho} u(z)}{\left[\ln\left(\frac{z-d}{z_M}\right) + \Psi_M\right]\left[\ln\left(\frac{z-d}{z_H}\right) + \Psi_H\right]}. \qquad (7.28)$$

Clearly the conductance depends on the height of the top of the air layer being considered.

Many of the computations performed simply ignore the profile diabatic correction factors (they are zero for neutral stability). When they are important, they are a bit of a challenge to compute, since the fluxes depend on the correction factors, but the correction factors depend on the heat flux density. One has to use an iterative approach and usually a computer to obtain a result. To give some idea of the importance of the stability correction, Fig. 7.2 shows atmospheric conductance for a range of wind speeds and canopy–air temperature differences. It can be seen that for stable conditions (canopy cooler than air) and low wind speed, the stability effect can be very large. At high wind speeds there is a smaller effect,

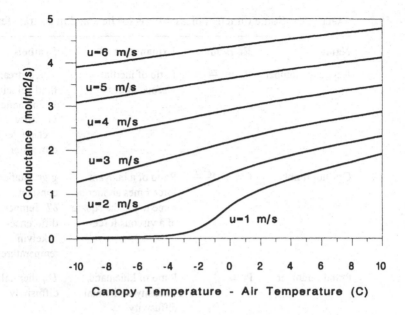

FIGURE 7.2. Effects of stability and wind on atmospheric boundary layer conductance. Measurement height is assumed to be twice the height of canopy, so if $z = 2$ m, then $h = 1$ m, $d = 0.65$ m, and $z_m = 0.1$ m.

and the canopy also remains closer to air temperature because of the high conductance. It is usually assumed that stability corrections are not important for wind speeds greater than 3 to 4 m/s. However, for low wind speeds during the night they can be very important.

7.7 Conductances for Heat and Mass Transfer in Laminar Forced Convection

In the first part of this chapter transfer of heat and mass by molecular diffusion is discussed. The fluid was assumed still or moving with laminar flow over an infinitely long surface so that there were no concentration gradients in the direction of flow. We now consider convective transport to or from a small object such as an animal's body or a leaf that is immersed in a fluid such as air. The assumptions made for molecular diffusion do not apply, but Eqs. (6.7) and (6.8) still apply if the conductances or resistances are properly defined. Our task here is to relate these terms to properties of the fluid and the surface. The spatial scale discussed here is between the scale of molecular diffusion (<mm) and the atmospheric surface layer (m to km), and ranges typically from mm to m.

A fundamental analysis of convective transport is extremely complicated and has not been accomplished for many surface shapes. For this reason an empirical approach is used. To make the results of empirical studies apply to as many different situations as possible, dimensionless

TABLE 7.3.   Some dimensionless groups for heat and mass transfer.

Name	Equation	Explanation	Symbols
Reynolds number	$Re = \frac{ud}{\nu}$	Ratio of inertial viscous forces	u free stream fluid velocity ν kinematic viscosity d characteristic dimension
Grashof number	$Gr = \frac{gd^3 \delta T}{T \nu^2}$	Ratio of a buoyant force times an inertial force to the the square of a viscous force	g gravitational constant δT Temperature difference T kelvin temperature
Prandtl number	$Pr = \frac{\nu}{D_H}$	Ratio of kinematic viscosity to thermal diffusivity	D_H thermal diffusivity
Schmidt number	$Sc = \frac{\nu}{D_j}$	Ratio of kinematic viscosity to mass diffusivity	D_j molecular diffusivity of species j
Nusselt number	$Nu = \frac{g_H d}{\rho D_H}$	Dimensionless conductance	ρ molar density g_H boundary layer conductance

groups of the variables have been formed and correlated empirically. By using these dimensionless groups, the results of a study on, say, heat transfer to water by a rod with a diameter of 1 cm, can be used to calculate heat transfer from a person's arm to air. It is therefore convenient to relate conductances to the appropriate dimensionless groups. Once this is done, the relationships between the dimensionless groups for describing the transport processes can be obtained directly from the engineering literature. The dimensionless groups that we use are given in Table 7.3. The Reynolds number, besides being useful for correlation of data on heat and mass transport, also gives an indication of whether the flow is laminar or turbulent. At low Re viscous forces predominate and the flow is laminar. At high Re, inertial forces predominate and the flow becomes turbulent. The critical Re at which turbulence starts is around 5×10^5 for a smooth, flat plate under "average" conditions.

Forced convection refers to the condition in which a fluid is moved past a surface by some external force (the analysis would be the same for a surface moving through a stationary fluid). Free convection, on the other hand, refers to fluid motion brought about by density gradients in the fluid as it is heated or cooled by the exchange surface. The boundary layer heat conductance for one surface (or "face") of a rectangular, flat plate with length d (note that the same symbol is used for the characteristic dimension in convection and the zero plane displacement in turbulent transport; do not confuse them), is

$$g_H = \frac{0.664 \hat{\rho} D_H \, Re^{1/2} \, Pr^{1/3}}{d}. \tag{7.29}$$

This equation is valid for any fluid, and requires only the correct viscosity, density, and diffusivity. If the Reynolds and Prandtl numbers are independent of pressure and temperature, then Eq. (7.29) has no pressure dependence, and the same temperature dependence as molecular diffusion. The fluid we are most interested in is air. Using the values from Table A.2 for air, Eq. (7.29) becomes

$$g_{Ha} = 0.135 \sqrt{\frac{u}{d}} \quad \left\{ \frac{mol}{m^2 s} \right\}. \tag{7.30}$$

The resistance of the boundary layer for heat transfer is the reciprocal of Eq. (7.30)

$$r_{Ha} = 7.4 \sqrt{\frac{d}{u}} \quad \left\{ \frac{m^2 s}{mol} \right\}. \tag{7.31}$$

The resistance for mass transfer is similar. For any fluid the relationship is:

$$g_j = \frac{0.664 \hat{\rho} D_j \, Re^{1/2} Sc^{1/3}}{d}. \tag{7.32}$$

The conductances in air for vapor, carbon dioxide, and oxygen are:

$$g_{va} = 0.147 \sqrt{\frac{u}{d}}$$

$$g_{ca} = 0.110 \sqrt{\frac{u}{d}} \quad \left\{ \frac{mol}{m^2 s} \right\} \tag{7.33}$$

$$g_{oa} = 0.130 \sqrt{\frac{u}{d}}.$$

The corresponding resistances are:

$$r_{va} = 6.8 \sqrt{\frac{d}{u}}$$

$$r_{ca} = 9.1 \sqrt{\frac{d}{u}} \quad \left\{ \frac{m^2 s}{mol} \right\} \tag{7.34}$$

$$r_{oa} = 7.7 \sqrt{\frac{d}{u}}.$$

7.8 Cylinders, Spheres and Animal Shapes

The relationships for conductance and resistance that were just presented can be derived from fundamental principles, but they apply only for transport from one side of a rectangular plate. With suitable adjustments in d, these could be used for leaves, but they would not necessarily apply for other surface shapes. So far it has not been possible to derive similar relationships from fundamental principles for objects like cylinders and spheres, so these relationships have been obtained empirically. Monteith and Unsworth (1990) and Simonson (1975) give examples of these relationships which are typical of those found in books on heat transfer. Rather than present these relationships here, we have plotted the ratio of the boundary layer conductance for a cylinder or a sphere to that for a rectangular plate. These ratios are shown in Fig. 7.3. Note that for a wide range of Reynolds numbers the ratio is within ±20 percent of unity. The Reynolds numbers of animals, fruits, etc. in outdoor wind are typically in the range shown in Fig. 7.3. Because of free stream turbulence in the atmosphere and other uncertainties, a 20 percent uncertainty in boundary layer conductance often is not bad. We therefore use the flat plate equation for all shapes of object. For improved estimates we can compute a Reynolds number and obtain a correction factor from Fig. 7.3. The correct conductance is just the flat plate conductance multiplied by the ratio from the figure. The characteristic dimension for computing the Reynolds

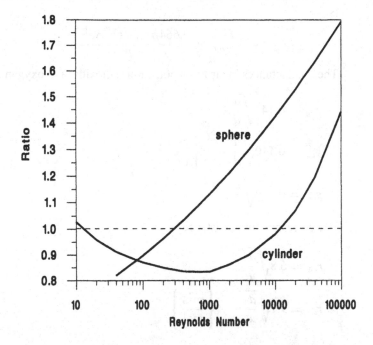

FIGURE 7.3. Ratio of cylinder or sphere conductance to plate conductance for a range of Reynolds numbers.

number or for computing the conductance is the diameter of the cylinder or the sphere. If the wind is blowing parallel to the axis of the cylinder, then it acts just like a flat plate, and the characteristic dimension is its length.

Another approach was taken by Mitchell (1976). When comparing measured conductances for animal shapes with the conductance of a sphere, it was found that conductances of shapes and sizes ranging from spiders and insects to cows were well represented by the relationship

$$g_{Ha} = \frac{0.34\hat{\rho}D_H \, \mathrm{Re}^{0.6}}{d} = 240u^{0.6}d^{-0.4}. \qquad (7.30a)$$

The characteristic dimension used by Mitchell for this calculation is the cube root of the volume of the animal. Equation (7.30a) is preferable to Eq. (7.30) except when the animal shape actually approximates a cylinder.

7.9 Conductances in Free Convection

Transport by free convection occurs whenever a body at one temperature is placed in a fluid at a higher or lower temperature. The heat transfer between the body and the fluid causes density gradients in the fluid, and these density gradients cause the fluid to mix. The transfer processes are similar to those in forced convection except the fluid velocity monatonically increases with distance from the surface in forced convection, but it first increases and then decreases to zero in free convection.

For laminar free convection, an expression for conductance that is adequate for cylinders (horizontal or vertical), spheres, vertical flat plates, and heated flat surfaces facing up or cooled surfaces facing down is:

$$g_H = \frac{0.54\hat{\rho}D_H(Gr \, \mathrm{Pr})^{1/4}}{d}. \qquad (7.35)$$

For cooled flat surfaces facing up or heated surfaces facing down the heat transfer is only about half as efficient so the constant 0.54 becomes 0.26. Again, this applies for any fluid. Substituting for the properties of air we obtain:

$$g_{Ha} = 0.05\left(\frac{T_s - T_a}{d}\right)^{1/4} \quad \left\{\frac{\mathrm{mol}}{\mathrm{m^2s}}\right\}. \qquad (7.36)$$

The resistance is just the reciprocal of Eq. (7.36).

The resistances for mass transfer by free convection are obtained by substituting the appropriate diffusivity and Schmidt number in Eq. (7.35). The free convection conductances for water vapor, CO_2 and O_2 are 1.09, 0.75, and 0.95 times the heat transfer conductance (Eq. 7.35).

Example 7.4. Find the flux densities of heat, CO_2, water vapor, and oxygen for a leaf under the conditions shown in the following table.

	Air	Leaf
Temperature	20° C	30° C
wind speed	3 m/s	
vapor concentration	15 mmol/mol	
carbon dioxide concentration	350 μmol/mol	70 μmol/mol
oxygen concentration	210.0 mmol/mol	210.2 mmol/mol

Assume the leaf has a characteristic dimension in the direction of wind flow of $d = 0.042$ m and that surface conductances (stomatal) to water vapor, CO_2, and O_2 transfer are 0.2, 0.13, and 0.18 mol m^2 s^{-1}.

Solution. The water vapor concentration at the leaf surface is the saturation concentration at leaf temperature. From Table A.3 the vapor pressure is 4.2 kPa, so the mole fraction (at 101 kPa) is 43 mmol/mol. The Reynolds number is Re = ud/ν = 3 m/s × 0.042 m/1.51 × 10^{-5} = 8.3 × 10^3. The flow apparently is laminar. All of the conductances except heat involve two conductors in series, the stomatal and the boundary layer. The conductances all involve:

$$\sqrt{\frac{u}{d}} = \sqrt{\frac{3}{0.042}} = 8.47.$$

The conductances are:

$$g_H = 0.135 \times 8.47 = 1.14 \text{ mol m}^{-2}\text{s}^{-1}$$

$$g_v = \frac{1}{\frac{1}{0.147 \times 8.47} + \frac{1}{0.2}} = 0.172 \text{ mol m}^{-2}\text{s}^{-1}$$

$$g_c = \frac{1}{\frac{1}{0.11 \times 8.47} + \frac{1}{0.13}} = 0.114 \text{ mol m}^{-2}\text{s}^{-1}$$

$$g_o = \frac{1}{\frac{1}{0.13 \times 8.47} + \frac{1}{0.18}} = 0.155 \text{ mol m}^{-2}\text{s}^{-1}.$$

The fluxes are:

$$H = c_p g_H (T_s - T_a) = 1.14 \frac{\text{mol}}{\text{m}^2\text{s}} \times 29.3 \frac{\text{J}}{\text{mol C}} (30°C - 20°C)$$

$$= 334 \frac{\text{W}}{\text{m}^2}$$

$$E = g_v (C_{vs} - C_{va}) = 0.172 \frac{\text{mol}}{\text{m}^2\text{s}} \times (0.042 - 0.015)$$

$$= 4.6 \frac{\text{mmol}}{\text{m}^2\text{s}}$$

$$F_c = g_c (C_{cs} - C_{ca}) = 0.114 \frac{\text{mol}}{\text{m}^2\text{s}} (70 \times 10^{-6} - 350 \times 10^{-6})$$

$$= -32 \frac{\mu\text{mol}}{m^2 s}$$

$$F_o = g_o(C_{os} - C_{oa}) = 0.155 \frac{\text{mol}}{m^2 s} \times (0.2102 - 0.2100)$$

$$= 31 \frac{\mu\text{mol}}{m^2 s}.$$

The negative sign on the carbon dioxide flux means that CO_2 is being taken up by the leaf. The fact that the oxygen and CO_2 fluxes are nearly equal in size and opposite in sign is the result of the stoichiometry of the photosynthesis reaction. One oxygen molecule is produced per CO_2 molecule.

7.10 Combined Forced and Free Convection

Almost all convective heat transfer processes in nature involve both forced and free convection. Usually one or the other process dominates and the conductance used is that calculated for the dominant process. The criterion normally used to determine which process is dominant is to determine the ratio Gr/Re^2. If this ratio is small, forced convection dominates. When the ratio is large, the opposite is true. When the ratio is near one, both forced and free convection must be considered, and the value of the resistance depends on whether the direction of the forced convection flow is such that it enhances or diminishes free convection. Additional detail on the mixed regime can be found in Kreith (1965).

Example 7.5. Find Gr for the previous example, and compute the ratio Gr/Re^2.

Solution. From Table 7.3:

$$Gr = \frac{gd^3(T_s - Ta)}{Tv^2} = \frac{9.8 \frac{m}{s^2} \times (0.042 \text{ m})^3 \times (30 \text{ C} - 20 \text{ C})}{298 \text{ K} \times \left(1.51 \times 10^{-5} \frac{m^2}{s}\right)^2}$$

$$= 1.1 \times 10^5$$

$$\frac{Gr}{Re^2} = \frac{1.1 \times 10^5}{(8.4 \times 10^3)^2} = 1.5 \times 10^{-3}.$$

Forced convection is obviously dominant.

7.11 Conductance Ratios

Situations often arise when the conductance for one species is known and we need to know the conductance for another. For example, we can easily determine the stomatal conductance for water vapor, but cannot determine it for CO_2. Is it possible to calculate one conductance if another is known? Considering any of the molecular diffusion equations (Eqs. (7.4) through (7.8)), the ratio of the conductances is equal to the ratio of the diffusivities

TABLE 7.4. Conductance ratios for
carbon dioxide and water vapor.

Process	Ratio	g_c/g_v
molecular diffusion	$(D_c/D_v)^1$	0.66
free convection	$(D_c/D_v)^{3/4}$	0.73
forced convection	$(D_c/D_v)^{2/3}$	0.75
turbulent transport	$(D_c/D_v)^0$	1.0

of the diffusing species. For free convection the ratios of the mass transport
equations which are similar to Eq. (7.35) are taken. In addition to the ratios
of the diffusivities, there are also the ratios of the Schmidt numbers. The
result is that the conductance ratio is equal to the diffusivity ratio to
the 3/4 power. Similarly, taking ratios using Eq. (7.32), and expanding
the Schmidt numbers gives the conductance ratio for forced convection.
The power of the diffusivities is now 2/3. Finally, for turbulent transport
Eq. (7.28) is used. There is no diffusivity dependence in these equations,
so the power is zero. These facts are summarized in Table 7.4, and values
for the ratio of CO_2 to water vapor conductance are given.

This exercise has produced a set of useful numbers, but it has not
given much explanation for why the conductance ratios differ for the
different processes. To get a little more insight into this, think of each
process as involving both diffusive and convective (meaning transport by
a moving fluid) transport. Differences in the size of molecules is important
in pure diffusion, and CO_2 diffuses much slower than water vapor. As
more and more of the transport occurs through fluid motion, the size of
the molecules has less and less effect. In turbulent transport there is no
effect. From Table 7.4 we see that even forced convection is still strongly
dominated by diffusive processes at the surface.

7.12 Determining the Characteristic Dimension of an Object

For a rectangular plate, the characteristic dimension is the length of the
plate in the direction the fluid is flowing. For a cylinder with its axis
parallel to the wind the characteristic dimension is also its length. The
characteristic dimension for cylinders and spheres is their diameter. A
characteristic dimension for various animal shapes can be obtained by
taking the cube root of the volume. For circular disks and various leaf
shapes, the characteristic dimension is more difficult to determine since
the width varies with distance along the leaf. The leaf can be divided into
a large number of rectangular pieces, each with its own characteristic
dimension, and these can be summed, with appropriate weighting, to
give a characteristic dimension in terms of a measurable dimension of

TABLE 7.5. Calculation of characteristic dimension for various shapes.

Object	Diagram	Explanation
Rectangular plate or cylinder with axis parallel to wind	wind $\longrightarrow$ d	d is the length of the plate or cylinder in the direction of flow
Circular disk	wind $\longrightarrow$ w	$d = 0.81w$, where w is the disk diameter.
Intersecting parabolas (leaf shape)	wind $\longrightarrow$ w	$d = 0.72w$, where w is the maximum leaf width in the direction of wind flow.
Sphere or cylinder (axis perpendicular to wind)	wind $\longrightarrow$ d	d is the diameter of the cylinder or sphere.
Animal shape		$d = V^{1/3}$, V is animal volume.

the leaf. For forced convection the characteristic dimension is computed from:

$$d = \left\{ \frac{\int_o^l d(y)\,dy}{\int_o^l \sqrt{d(y)}\,dy} \right\}^2 \tag{7.37}$$

where l is the length of the leaf perpendicular to the wind and $d(y)$ indicates the variation of the leaf width with distance along the length of the leaf. Table 7.5 gives values for a circular disk and a leaf shape (intersecting parabolas) from Eq. (7.37). Leaves, of course, have many different shapes, and none of them exactly matches the overlapping parabolas in Table 7.5. Typically, however, a factor around 0.7 converts maximum dimension in the direction of wind flow to characteristic dimension. This is also a typical ratio of the area of a leaf to the product of its length and width.

7.13 Free Stream Turbulence

Most of the relationships presented so far describe conductances one would measure in a carefully constructed wind tunnel which minimizes the turbulence in the air. When turbulence is induced in the air that flows over the object, the conductance increases, often dramatically. Conductances in wind tunnels with turbulence generated by placing obstructions upwind of the object are sometimes twice those for laminar flow.

The outdoor wind is naturally turbulent, so the conductance of objects placed in natural wind is likely to be higher than would be predicted using the equations presented so far. The size of this enhancement is determined by the size of the object and the size of the eddies in the air. The eddy size increases with height in the atmosphere, so the enhancement should change with height. Mitchell (1976) measured convective heat transfer from spheres in the atmosphere and related the enhancement to the ratio of sphere diameter to distance above the ground. Mitchell's relation is shown in Fig. 7.4. For heights ranging from about 2 to 10 times the object diameter the enhancement is around 1.4, and this is the value we use for all outdoor computations.

Summary of Formulae for Conductance

The formulae given in this chapter are used frequently throughout the rest of the book. It is therefore convenient to summarize them in a single location. This is done in Table 7.6.

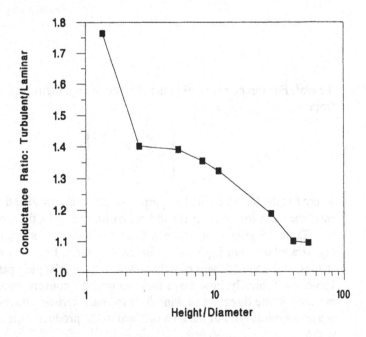

FIGURE 7.4. Enhancement of conductance by free stream turbulence for spheres at various heights in natural wind. (from Mitchell, 1976).

TABLE 7.6. Formulae for calculating conductances (mol m^{-2} s^{-1}) for diffusion, convection, and turbulent transport in air.

Process	Heat Transfer	Vapor Transfer	Other Mass Transfer
conduction or diffusion (molecular processes)	Plane $g_H = \frac{\hat{\rho}D_H}{\Delta z}$ Sphere $g_H = \frac{\hat{\rho}\,3D_H}{z_s\left(1-\frac{z_s}{z_a}\right)}$ Cylinder $g_H = \frac{\hat{\rho}D_H}{z_s \ln\frac{z_a}{z_s}}$	Plane $g_v = \frac{\hat{\rho}D_v}{\Delta z}$ Sphere $g_v = \frac{\hat{\rho}D_v}{z_s\left(1-\frac{z_s}{z_a}\right)}$ Cylinder $g_v = \frac{\hat{\rho}D_v}{z_s \ln\frac{z_a}{z_s}}$	Plane $g_j = \frac{\hat{\rho}D_j}{\Delta z}$ Sphere $g_j = \frac{\hat{\rho}D_j}{z_s\left(1-\frac{z_s}{z_a}\right)}$ Cylinder $g_j = \frac{\hat{\rho}D_j}{z_s \ln\frac{z_a}{z_s}}$
Forced convection (fluid moved past surface by external force)	$g_{Ha} = 0.135\sqrt{\frac{u}{d}}$	$g_{va} = 0.147\sqrt{\frac{u}{d}}$	$g_{ca} = 0.110\sqrt{\frac{u}{d}}$ $g_{oa} = 0.130\sqrt{\frac{u}{d}}$
Free convection (fluid flow generated from temperature gradients)	$g_{Ha} = 0.050\left(\frac{T_s-T_a}{d}\right)^{1/4}$	$g_{va} = 0.055\left(\frac{T_s-T_a}{d}\right)^{1/4}$	$g_{ca} = 0.038\left(\frac{T_s-T_a}{d}\right)^{1/4}$ $g_{oa} = 0.048\left(\frac{T_s-T_a}{d}\right)^{1/4}$
Eddy diffusion or turbulent transport (wind over fields)	$g_{Ha} = g_{va} = g_{ca} = \dfrac{0.4^2\hat{\rho}u(z)}{\left[\ln\left(\frac{z-d}{z_M}\right)+\psi_M\right]\left[\ln\left(\frac{z-d}{z_H}\right)+\psi_H\right]}$		

References

Cowan, I.R. (1972) Mass and heat transfer in laminar boundary layers with particular reference to assimilation and transpiration in leaves. Agric. Meteor. 10:311–329.

deMichael, D.W. and P.J.H. Sharp (1973) An analysis of the mechanics of guard cell motion. J. Theor. Biol. 41:77–96.

Eckert, E.R.G. and R.M. Drake (1972) Analysis of Heat and Mass Transfer New York: McGraw-Hill.

Garrat, J.R., and B.B. Hicks (1973) Momentum, heat, and water vapor transfer to and from natural and artificial surfaces. Quart. J. Roy. Meteor. Soc. 99:680–687.

Jarman, P.D. (1974) The diffusion of carbon dioxide and water vapor through stomata. J. Exp. Bot. 25:927–936

Kowalski, G.J. and J.W. Mitchell (1975) Heat transfer from spheres in the naturally turbulent, outdoor environment. Amer. Soc. Mech. Eng. Paper No. 75-WA/HT-57

Kreith, F. (1965) Principles of Heat Transfer. Scranton, Pa.: International Textbook Co.

Lumley, J.L., and H.A. Panofsky (1964) The structure of Atmospheric Turbulence. New York: Wiley.

Mitchell, J.W. (1976). Heat transfer from spheres and other animal forms. Biophysical Journal 16:561–569.

Monteith, J.L., and G.S. Campbell (1980) Diffusion of water vapour through integuments—potential confusion. J. Therm. Biol. 5:7–9.

Monteith, J.L. and M.H. Unsworth (1990) Principles of Environmental Physics. 2nd ed. London: Edward Arnold.

Nobel, P.S. (1974) Boundary layers of air adjacent to cylinders Plant Physiol. 54:177–181.

Parkhurst, D.F., P.R. Duncan, D.M. Gates, and F. Kreith (1968) Wind tunnel modeling of convection of heat between air and broad leaves of plants Agric. Meteor. 5:33.

Patten, K.D., G.W Apel and M.E. Patterson (1988) Cuticular diffusive resistance measurements: A technique for teaching and research in postharvest research in horticulture. Dept. of Horticulture, Wash. State Univ., Pullman, WA.

Yasuda, N. (1988) Turbulent diffusivity and diurnal variations in the atmospheric boundary layer. Boundary-Layer Meteorol. 43:209–221.

Problems

7.1. Maximum width of a leaf in the direction of wind flow is 5 cm. Leaf temperature is 20° C in a 1 m/s wind when $T_a = 15°$ C. Find d, Re, Gr, g_{Ha}, and H. Is heat transfer mainly by forced or free convection?

7.2. The wind speed at a height of 2 m is 5.6 m/s. Find the boundary layer (turbulent transport) conductance for a potato canopy that is 50 cm high. Assume neutral stability.

7.3. Find the heat flux density from your arm, at 33° C, to room air at 22° C if there is no wind (free convection) and if the wind speed is 2 m/s.

7.4. Compute the heat loss from a sheep to the air if its fleece is 5 cm thick and the diameter of the sheep's body (inside the fleece) is 25 cm. Assume the fleece has twice the conductance of still air, the body temperature is 37° C, the wind speed is 4 m/s, and the air temperature is 5° C.

7.5. Compare thermal conductance of clothing at sea level with conductance at 5000 m elevation. Is elevation likely to have a noticeable effect on heat loss through clothing?

7.6. In Eq. (7.10) there is no minus sign as in Eq. (7.9). Based on the sign convention implied by Eq. (7.13), discuss the relationship between fluctuations of w and T and identify what conditions are associated with a flux from the soil surface to the atmosphere.

Heat Flow in the Soil 8

When the sun shines on the soil surface, some of the energy is absorbed, heating the soil surface. This heat is lost from the surface through conduction to lower layers of the soil, through heating the atmosphere, and through evaporation of water. Heat transport from the surface to the atmosphere was discussed in Ch. 7. This chapter considers heat transport into the soil. Some of the results from an analysis of heat transport in soil are presented in Ch. 2 to show typical temporal and spatial patterns of soil temperature. Here we show how those equations are derived and how they depend on soil properties.

8.1 Heat Flow and Storage in Soil

In analyzing heat flow in the soil or the atmosphere, it is useful to mentally divide the medium into a large number of thin layers, and consider the heat flow and storage in each layer. The amount of heat stored in a layer of air is small compared to the amount of heat transferred through it. Within the first few meters of the atmosphere the heat stored in the air is generally ignored and heat transfer processes are assumed to be approximately steady. The results of these assumptions are the equations developed in Ch. 7.

In soil the storage term is much larger and cannot be ignored. The heat flow from one layer to the next is still computed using the Fourier law (Eq. (6.3)) but now the continuity equation must be solved simultaneously to find the temperature variation with depth and time. The continuity equation is:

$$\rho_s c_s \frac{\partial T}{\partial t} = -\frac{\partial G}{\partial z} \tag{8.1}$$

where ρ_s is the density of the soil, c_s is the soil specific heat, $\rho_s c_s$ is the volumetric heat capacity, and G is the heat flux density in the soil (from Eq. (6.3)). The left-hand side of Eq. (8.1) represents the rate of heat storage in a layer of soil and the right-hand side represents the heat flux divergence, or rate of change of heat flux density with depth.

Combining eqs. 6.3 and 8.1 gives

$$\rho_s c_s \frac{\partial T}{\partial t} = \frac{\partial}{\partial z}\left(k\frac{\partial T}{\partial z}\right). \tag{8.2}$$

If thermal conductivity is constant with depth, k can be taken outside the derivative. We can also divide both sides by $\rho_s c_s$ to obtain a more familiar form of the heat equation:

$$\frac{\partial T}{\partial t} = \kappa\frac{\partial^2 T}{\partial z^2} \tag{8.3}$$

where

$$\kappa = \frac{k}{\rho_s c_s} \tag{8.4}$$

is the soil thermal diffusivity. According to Eq. (8.3), the location in the soil where temperature will change fastest with time is the location where the change with depth of the temperature *gradient* is largest.

In principle, solutions to Eq. (8.2) can simulate the behavior of soil temperature in space and time. The conditions for which analytic solutions can be obtained, however, are very restrictive, and do not represent real soil environments very well. Realistic conditions can be simulated by solving the equation numerically, but these solutions are not very useful for understanding the behavior of the system. We now look at a couple of simple solutions to Eq. (8.3). These are useful for understanding, at least qualitatively, spatial and temporal patterns in soil temperature.

If the soil is assumed to be infinitely deep, with uniform thermal properties, and a surface temperature that varies sinusoidally according to the equation:

$$T(0, t) = T_{ave} + A(0)\sin\left[\omega(t - t_o)\right] \tag{8.5}$$

then the temperature at any depth and time is given by:

$$T(z, t) = T_{ave} + A(0)\exp(-z/D)\sin\left[\omega(t - t_o) - z/D\right]. \tag{8.6}$$

where t_o is a phase shift that depends on whether t is local time, universal time or some other time reference. In Eq. (2.4) local time was used and $t_o = 8$. Recall from Ch. 2 that T_{ave} is the average temperature over a temperature cycle, $A(0)$ is the amplitude of the temperature fluctuations (half the difference between minimum and maximum) and ω is the angular frequency, which is calculated from

$$\omega = \frac{2\pi}{\tau} \tag{8.7}$$

where τ is the period of the temperature fluctuations. In Ch. 2 we were using time in hours, so τ was in hours, but here we need τ in seconds.

We are now interested in diurnal and annual fluctuations so

$$\omega_{diurnal} = \frac{2\pi}{24 \times 3600s} = 7.3 \times 10^{-5}\,s^{-1}$$

$$\omega_{annual} = \frac{2\pi}{365 \times 24 \times 3600\,s} = 2 \times 10^{-7}s^{-1}.$$

The symbol D represents the damping depth, and is calculated from:

$$D = \sqrt{\frac{2\kappa}{\omega}}. \qquad (8.8)$$

Referring to Eq. (8.6), it can be seen that D determines how much the amplitude of the temperature variation is attenuated with depth and how much the phase is shifted in time. When $z = D$ the exponential in Eq. (8.6) has a value of 0.37, indicating that the amplitude of temperature fluctuations at that depth is 37 percent of the amplitude at the surface. At $z = 2D$ the amplitude is $\exp(-2) = 0.14$, and at $z = 3D$ the amplitude is $\exp(-3) = 0.05$. The damping depth therefore gives useful information about the depth to which temperature fluctuations penetrate into the soil. Even though the surface temperature is not sinusoidal, the damping depth still gives a good idea of how deep diurnal and annual temperature fluctuations will penetrate.

The damping depth also affects the phase. At the depth where $z/D = \pi$, or $z = \pi D$, the temperature reaches a maximum when the surface temperature is at its minimum. To get an overall picture of temperature variation with depth and time Eq. (8.6) can be plotted in three dimensions. This is shown in Fig. 8.1. Note how the temperature fluctuations are attenuated with depth and are shifted in time. At the bottom of the graph, amplitude is only about five percent of the amplitude at the surface and the maximum occurs at about the same time as the minimum at the surface.

To find the heat flux density at the soil surface differentiate Eq. (8.6), substitute from Eq. (6.3), and set z to zero. Doing this gives

$$G_{(0,t)} = \frac{\sqrt{2}A(0)k \sin[\omega(t - t_0) + \pi/4]}{D}. \qquad (8.9)$$

Equation (8.9) shows that the maximum heat flux density occurs 1/8 cycle ($\pi/4$) before the maximum temperature (Eq. (8.6)). This flux can be integrated over a half-cycle to determine the total heat input to the soil. From the integration $\sqrt{2}D\rho_s c_s A(0)$ is obtained, which is the same as the heat storage that would occur in a layer of soil of thickness $\sqrt{2}D$ which changed temperature by $A(0)$. Therefore, $\sqrt{2}D$ can be thought of as an effective depth for thermal exchange with the soil.

Yet another relationship can be obtained from Eq. (8.9), or from the expression for total heat input. Using Eq. (8.8) the following can be written:

$$\frac{\sqrt{2}k}{D} = \mu\sqrt{\omega} \quad \text{and} \quad \sqrt{2}D\rho_s c_s = \frac{2\mu}{\sqrt{\omega}}$$

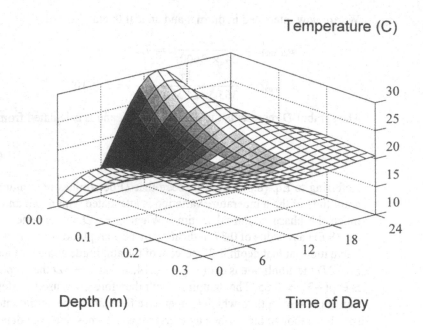

FIGURE 8.1. Graph of Eq. (8.6) showing how the surface temperature wave is attenuated with depth and shifted in time.

where $\mu = \sqrt{k\rho_s c_s}$. This square root of the product of thermal conductivity and volumetric heat capacity is called the thermal admittance, μ. It can be seen that this relates directly to the ability of the soil to store heat, since both the rate of heat storage (Eq. (8.9)) and the total amount of heat stored in a half-cycle are proportional to the thermal admittance. Soils with a high thermal admittance store heat more readily than those with low admittance. When the admittance is high much of the heat available at the surface goes to heating the soil, while when it is low, most of the heat goes to the atmosphere.

The thermal admittance can be used to help understand how radiant energy that is absorbed at a dry surface might be partitioned between the atmosphere (convection) and the soil (conduction). Since the soil surface is dry, it can be assumed that latent heat loss is near zero, so radiant energy is approximately equal to $G + H$. This is partitioned as:

$$\frac{G}{H} = \frac{\mu_{\text{soil}}}{\mu_{\text{atmos}}} \tag{8.10}$$

Some approximate values of G/H are given in Table 8.1 for a dry bare soil and a dry mulch.

This analysis is only qualitative because Eq. (8.10) assumes μ is constant with height, and the equations derived in Ch. 7 show that μ increases with height. However, it can be seen that a higher atmospheric admit-

TABLE 8.1. Dependence of soil and sensible heat flux on atmospheric and soil factors for two dry surfaces.

Medium	G/H	
	Bare, dry soil	Loose straw mulch
Still air	50	20
Calm atmosphere	0.5	0.2
Windy atmosphere	0.1	0.04

tance (windy atmosphere) or a lower soil admittance (loose straw mulch) decreases the heat going to the soil.

The thermal admittance can also be used to estimate the contact surface temperature at the interface between two solid objects, each initially at a different temperature, when they are brought into contact. If one object with an initial temperature T_1 and thermal admittance μ_1 is brought into contact with another object with initial temperature T_2 and thermal admittance μ_2, then the temperature at the interface, T_s, is given by

$$T_s = \frac{\mu_1 T_1 + \mu_2 T_2}{\mu_1 + \mu_2}. \qquad (8.11)$$

Clearly the object with the higher thermal admittance will dominate the interface temperature. This is why a tile floor "feels" colder than a carpet. The tile has a much higher admittance than the carpet.

8.2 Thermal Properties of Soils: Volumetric Heat Capacity

In order to compute damping depths, admittances, and soil temperature profiles, the thermal diffusivity of the soil needs to be known. This, in turn, requires a knowledge of the thermal conductivity and specific heat of the soil. In this section we tell how to find these quantities.

The volumetric heat capacity of a soil is the sum of the heat capacities of the soil components. Soil typically is made up of minerals, water, and organic matter. The soil heat capacity is therefore computed from

$$\rho_s c_s = \phi_m \rho_m c_m + \theta \rho_w c_w + \phi_o \rho_o c_o \qquad (8.12)$$

where θ is water content (volume fraction of water), ϕ_m and ϕ_o are volume fractions of minerals and organic material, and c and ρ are the specific heat and density. While air is almost always present, its contribution to the soil heat capacity is negligible. Other constituents, like ice, are added to Eq. (8.12) when present. Table 8.2 lists thermal properties for a number of soil constituents. Thermal properties with significant temperature dependence are indicated.

Figure 8.2 shows the variation in heat capacity of four typical soils when the water content varies from zero to saturation. As indicated by Eq. (8.12), the change is linear, and values range from less than 0.5 to

TABLE 8.2. Thermal properties of typical soil materials.

Material	Density (Mg m^{-3})	Specific Heat (J g^{-1} K^{-1})	Thermal Conductivity (W m^{-1} K^{-1})	Volumetric heat capacity (MJ m^{-3} K^{-1})
Soil minerals	2.65	0.87	2.5	2.31
Granite	2.64	0.82	3.0	2.16
Quartz	2.66	0.80	8.8	2.13
Glass	2.71	0.84	0.8	2.28
Organic matter	1.30	1.92	0.25	2.50
Water	1.00	4.18	$0.56 + 0.0018T$	4.18
Ice	0.92	$2.1 + 0.0073T$	$2.22 - 0.011T$	$1.93 + .0067T$
Air (101 kPa)	$(1.29 - 0.0041T)$ $\times 10^{-3}$	1.01	$0.024 + 0.00007T$	$(1.3 - 0.0041T)$ $\times 10^{-3}$

about 3.5 MJ m^{-3} K^{-1}. The slope of all lines is the same and is determined by the heat capacity of water. The intercepts differ because of the differences in solid fractions in the different soils.

Example 8.1. Find the volumetric heat capacity of loam soil with a water content of 0.2 m^3 m^{-3} and a bulk density of 1.3 Mg/m^3. Assume the organic fraction is zero.

Solution. The mineral fraction is the ratio of the bulk density to the mineral density. The mineral density of soil, from Table 8.2, is 2.65 Mg/m^3,

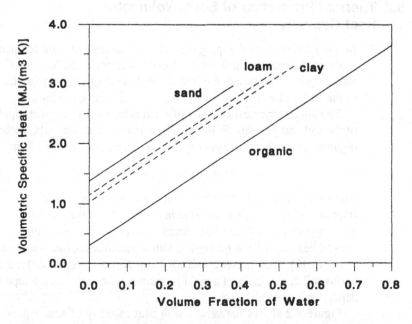

FIGURE 8.2. Volumetric heat capacity of organic and mineral soils. Differences are mainly due to differences in soil bulk density.

so $\phi_m = 1.3/2.65 = 0.49$. Using Eq. (8.12),

$$\rho_s c_s = 0.49 \times 2650 \frac{kg}{m^3} \times 870 \frac{J}{kg\ C}$$

$$+ 0.2 \times 1000 \frac{kg}{m^3} \times 4180 \frac{J}{kg\ C} = 1.97 \times 10^6 \frac{J}{m^3\ C}$$

8.3 Thermal Properties of Soils: Thermal Conductivity

The thermal conductivity of soil depends on the conductivities and volume fractions of the soil constituents. The heat flows through a complicated network of mineral, water, and air paths and the quantity and conductivity of each strongly influences the effectiveness of the others. In addition, a substantial quantity of heat is carried by evaporation and condensation in the soil pores, and this is both water content and temperature dependent. DeVries (1963) proposed that the thermal conductivity of soil be computed as a weighted sum of the conductivities of the constituents:

$$k_{soil} = \frac{\phi_w \xi_w k_w + \phi_g \xi_g k_g + \phi_m \xi_m k_m}{\phi_w \xi_w + \phi_g \xi_g + \phi_m \xi_m} \tag{8.13}$$

where ϕ is the volume fraction, ξ is a weighting factor, k is the thermal conductivity of the constituent, and subscripts w, g, and m indicate the water, gas, and mineral fractions.

The apparent thermal conductivity of the gas phase is the sum of the thermal conductivity of air, given in Table 8.2, and an apparent conductivity resulting from latent heat transport within the pores of the soil. Water evaporates on one side of the pore, diffuses across the pore in the air space, and then condenses on the other side of the pore. The latent heat of evaporation is carried with the water across the pore. After the water condenses, it can flow back to the hot side of the pore and evaporate again. Engineers have used this same idea in highly effective heat exchangers called heat pipes. The pipes are tubes with a volatile liquid and a wick sealed inside. The liquid evaporates on the hot end of the tube, diffuses to the cold end, condenses, and then moves back to the hot end through the wick. The heat pipe is sealed so there is always plenty of liquid, but the soil can dry out. As the soil water content decreases, the water films become thinner, and the return flow of liquid water in the soil pores is increasingly impeded until there is no contribution of latent heat to the overall heat transport in soil pores.

Fick's law can be used to compute the latent heat flow in a pore. Using Eq. (6.5) gives:

$$\lambda E = -\lambda \hat{\rho} D_v \frac{dC_v}{dz} = -\lambda \hat{\rho} D_v \frac{dC_v}{dT} \frac{dT}{dz} \tag{8.14}$$

where $\hat{\rho}$ is the molar density of air, λ is the latent heat of vaporization of water, D_v is the vapor diffusivity for soil, and C_v is the vapor mole fraction given by the ratio of vapor pressure divided by total atmospheric pressure (e/p_a). The second equation is obtained by applying the chain rule of calculus. The derivative of water concentration with respect to temperature can be expanded using the relationship $C_v = h_r C_s(T)$ from Ch. 3 where h_r is the relative humidity in the soil. Since h_r is not temperature dependent, it can be taken out of the derivative. Now, using another definition from Ch. 3: $s = dC_s(T)/dT$, gives the slope of the saturation mole fraction function for water; which is simply related to the slope of the saturation vapor pressure versus temperature. Substituting these into Eq. (8.14) gives

$$\lambda E = -\lambda \hat{\rho} D_v h_r s \frac{dT}{dz}. \tag{8.15}$$

The apparent thermal conductivity for distillation across a pore is made of all the terms which multiply the temperature gradient.

Equation (8.15) is adequate for moist soils at low temperature, but requires two corrections for it to work at high temperatures or for dry soils. When water evaporates from a surface, mass in the vapor phase is created at the liquid–gas interface which causes the entire gas phase to flow away from the surface. At low temperature this mass flow effect is negligible, but at boiling point its effect is far greater than the diffusive flux from Fick's law. The correction to the equation is called the Stefan correction. It can be inserted into Eq. (8.15) by substituting $\Delta/(p_a - e_a)$ for s where Δ is the slope of the saturation vapor pressure function. From Ch. 3, $s = \Delta/p_a$. At typical environmental temperatures $p_a \gg e_a$ this substitution will have very little effect. If a moist soil is heated by a fire at the surface, however, the Stefan correction becomes very large, and the soil becomes an excellent conductor of heat because $p_a - e_a$ becomes small:

$$\lambda E = -\lambda \hat{\rho} D_v h_r \frac{\Delta}{p_a - e_a} \frac{dT}{dz}. \tag{8.16}$$

The second correction was mentioned previously relating to the return flow of water. Even before the humidity in the soil drops significantly below one (remember from Ch. 4 that the humidity in moist soil is always close to one) the return flow of liquid water in the soil pores has dropped sufficiently to render the latent heat component of the pore conductivity negligible. No fundamental theory has been developed yet to account for this. Campbell et al. (1994) give a dimensionless flow factor which depends on the soil water content. This factor multiplies Eq. (8.16) to give the actual latent heat flux. The factor is

$$f_w = \frac{1}{1 + \left(\frac{\theta}{\theta_o}\right)^{-q}}. \tag{8.17}$$

The constant θ_o determines the water content where return flow cuts off and q determines how quickly the cutoff occurs. Both constants are correlated with soil texture and tend to increase as textures become finer. The range for θ_o is from around 0.05 for coarse sand to 0.25 for heavy clay. The range for q is roughly 2 to 6 with coarser materials generally having lower values, but the pattern is not as clear as for the cutoff water content.

The complete expression for vapor phase apparent conductivity is

$$k_g = k_a + \frac{\lambda \Delta h_r f_w \hat{\rho} D_v}{p_a - e_a} \tag{8.18}$$

where k_a is the thermal conductivity of air. The weighting factors are determined by the shapes, conductivities, and volume fractions of the soil constituents. Campbell et al. (1994) defined a fluid conductivity for the soil as

$$k_f = k_g + f_w(k_w - k_g). \tag{8.19}$$

In dry soil the fluid conductivity is the value for dry air and in saturated soil it is the value for water. The same f_w function used for the liquid return flow is used in Eq. (8.19). Using Eq. (8.19) the weighting functions can now be computed:

$$\begin{aligned}
\xi_g &= \frac{2}{3[1 + g_a(k_g/k_f - 1)]} + \frac{1}{3[1 + g_c(k_g/k_f - 1)]} \\
\xi_w &= \frac{2}{3[1 + g_a(k_w/k_f - 1)]} + \frac{1}{3[1 + g_c(k_w/k_f - 1)]} \\
\xi_m &= \frac{2}{3[1 + g_a(k_m/k_f - 1)]} + \frac{1}{3[1 + g_c(k_m/k_f - 1)]}.
\end{aligned} \tag{8.20}$$

In these equations, k_g is from Eq. (8.18), while k_w and k_m are from Table 8.2. The shape factors, g_a and g_c, depend on the shape of the soil particles. One can compute g_c from $g_c = 1 - 2g_a$. For mineral soils g_a has a value around 0.1. For organic soils it is 0.33.

These equations are most useful as part of a computer program as they are quite long for hand calculations. They do, however, include all of the effects of temperature, moisture, density, and soil composition. The interaction among these factors is complex and significant, and, at present, no simpler approach is apparent. Figure 8.3 shows thermal conductivity computed using Eq. (8.13) for the soils in Fig. 8.2. The mineral conductivities of the clay and loam samples are 2.3 and 2 W m^{-1} C^{-1}. The organic k_m is 0.3 and the sand k_m is 5 W m^{-1} C^{-1}. The sand curve is meant to represent a sample with high quartz content.

Example 8.2. For the soil in Example 8.1, find the thermal conductivity. Assume $T = 20°$ C, $p_a = 101$ kPa, $k_m = 2.5$ W m^{-1} C^{-1}, $q = 4$, and $\theta_o = 0.15$.

Solution. To do the calculation, values are needed for λ, Δ, h_r, D_v, and e. From Table A.2, $\lambda = 44100$ J/mol. From Table A.3, $\Delta = 145$ Pa/C and $e = 2340$ Pa. From Table A.1, $D_v = 2.42 \times 10^{-5}$ m^2/s. From Table 8.2, $k_a = 0.025$ W m^{-1} C^{-1} and $k_w = 0.60$ W m^{-1} C^{-1}. There is no easy way of calculating h_r, but a loam soil at 0.2 water content is well above

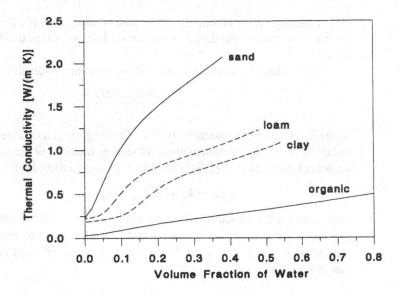

FIGURE 8.3. Thermal conductivity of mineral and organic soils from Eq. (8.11). Temperature is 20° C.

permanent wilting point, so the humidity must be nearly 1.0. The volume fraction of minerals is 0.49, from example 1, and the water fraction is 0.2. The gas fraction is $\phi_g = 1 - \theta - \phi_m = 1 - 0.2 - 0.49 = 0.31$.

From Eq. (8.17), the dimensionless flow factor is

$$f_w = \frac{1}{1 + \left(\frac{0.2}{0.15}\right)^{-4}} = 0.76.$$

The thermal conductivity of the gas phase is (Eq. (8.18)):

$$k_g = 0.025$$
$$+ \frac{44100\,\frac{J}{mol} \times 145\,\frac{Pa}{C} \times 1.0 \times 0.76 \times 41.4\,\frac{mol}{m^3} \times 2.42 \times 10^{-5}\,\frac{m^2}{s}}{101000\,Pa - 2340\,Pa}$$
$$= 0.074\ \mathrm{W\ m^{-1}\ C^{-1}}$$

Note that the contribution from latent heat transport is about twice that for conduction through the dry air. Using Eq. (8.19) the fluid conductivity can be computed. It is

$$k_f = 0.074\,\frac{W}{mC} + 0.76 \times (0.6 - 0.074)\,\frac{W}{mC} = 0.47\,\frac{W}{mC}.$$

Now the weighting factors can be computed using Eq. (8.20):

$$\xi_g = \frac{2}{3\left[1 + 0.1\left(\frac{0.074}{0.47} - 1\right)\right]} + \frac{1}{3\left[1 + 0.8\left(\frac{0.074}{0.47} - 1\right)\right]} = 1.75$$

$$\xi_w = \frac{2}{3\left[1 + 0.1\left(\frac{0.6}{0.47} - 1\right)\right]} + \frac{1}{3\left[1 + 0.8\left(\frac{0.6}{0.47} - 1\right)\right]} = 0.92$$

$$\xi_m = \frac{2}{3\left[1 + 0.1\left(\frac{2.5}{0.47} - 1\right)\right]} + \frac{1}{3\left[1 + 0.8\left(\frac{2.5}{0.47} - 1\right)\right]} = 0.54$$

where we have assumed $g_a = 0.1$. Equation (8.13) is now used to find the thermal conductivity:

$$k_s = \frac{0.31 \times 1.75 \times 0.074 \frac{W}{mC} + 0.2 \times 0.92 \times 0.6 \frac{W}{mC} + 0.49 \times 0.54 \times 2.5 \frac{W}{mC}}{0.31 \times 1.75 + 0.2 \times 0.92 + 0.49 \times 0.54}$$

$$= 0.82 \frac{W}{mC}.$$

Even though the air has a very low thermal conductivity, it profoundly influences the conductivity of the soil when the gas fraction is high. Most of the heat has to flow through the air spaces, so they exert a controlling influence on overall heat flow. The model accounts for this through the fact that the weighting factor for the gas phase is larger than the other two factors.

The slope of the saturation vapor pressure function is strongly temperature dependent, so the apparent thermal conductivity of the gas phase increases rapidly with temperature. In the example just described, the gas phase conductivity is only a little over 10 percent of the water conductivity, but as temperature increases they become more similar. At about 60° C, the gas and water phase conductivities are equal, so for moist soil ($f_w \approx 1$), the conductivity becomes independent of water content.

8.4 Thermal Diffusivity and Admittance of Soils

Equation (8.4) defines the thermal diffusivity as the ratio of conductivity to volumetric heat capacity. Figure 8.4 shows the diffusivity for the soils in Figs. 8.2 and 8.3. The diffusivity of the organic soil is almost constant with water content, while the mineral soils have a relatively rapid transition from dry to wet diffusivity. The sand diffusivity is so much higher than the others mainly because we assumed a high quartz content for it. A sand with mineral conductivity equal to that for the loam and clay would have diffusivities near the loam line. We also assumed a higher bulk density for the sand, which also increased its diffusivity. A low-quartz soil with average bulk density would have a dry diffusivity around 0.2 mm²/s and a wet diffusivity around 0.4 mm²/s.

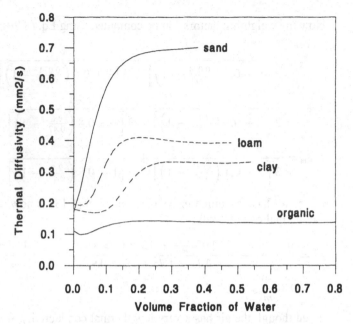

FIGURE 8.4. Thermal diffusivities of soils from Figs. 8.2 and 8.3.

Example 8.3. Compare the diurnal damping depths for moist organic soil with dry and wet loam soil.

Solution. From Fig. 8.4 the diffusivities appear to be around 0.14, 0.2, and 0.4 mm^2/s for the organic, dry loam, and wet loam respectively. Using Eq. (8.8),

$$D_{\text{organic}} = \sqrt{\frac{2 \times 0.14 \times 10^{-6} \frac{\text{m}^2}{\text{s}}}{7.3 \times 10^{-5} \text{ s}^{-1}}} = 0.062 \text{ m} = 6.2 \text{ cm}$$

$$D_{\text{dry loam}} = \sqrt{\frac{2 \times 0.2 \times 10^{-6} \frac{\text{m}^2}{\text{s}}}{7.3 \times 10^{-5} \text{ s}^{-1}}} = 0.074 \text{ m} = 7.4 \text{ cm}$$

$$D_{\text{wet loam}} = \sqrt{\frac{2 \times 0.4 \times 10^{-6} \frac{\text{m}^2}{\text{s}}}{7.3 \times 10^{-5} \text{ s}^{-1}}} = 0.105 \text{ m} = 10.5 \text{ cm}.$$

If the same diurnal temperature variation were applied to the surface of each profile, it would penetrate about twice as deep into the wet mineral soil as it would the organic soil.

Example 8.4. At what depth would soil temperature be measured if you wanted to find the mean annual temperature to within 1° C?

Solution. From Ch. 2 we know that the range of annual temperature variation is 20 to 30° C, so $A(0)$ is 10 to 15° C. We want the variation to be less than 1° C, so, using Eq. (8.6)

$$15\,C \times \exp\left(-\frac{z}{D}\right) = 1\,C.$$

We assume that the soil is moist for most of the year. The annual damping depth is therefore

$$D_{\text{annual}} = \sqrt{\frac{2 \times 0.4 \times 10^{-6}\,\frac{m^2}{s}}{2 \times 10^{-7}\,s^{-1}}} = 2\,m.$$

Solving for z gives

$$z = D_{\text{annual}}\ln(15) = 2m \times 2.7 = 5.4m.$$

The same type of calculation could be used to find the average temperature over a diurnal cycle. It would also be roughly three times the damping depth, or about 30 cm.

As previously discussed, the thermal admittance, or ability of the soil to store heat when temperature varies over a specified range, is the square root of the product of thermal conductivity and volumetric heat capacity. The information in Figs. 8.2 and 8.3 can be combined to give thermal admittance values. These are shown in Fig. 8.5. Water content affects the admittance of all of the soils dramatically over the whole range of water contents. Wet soil admittances are four to five times those of dry soils.

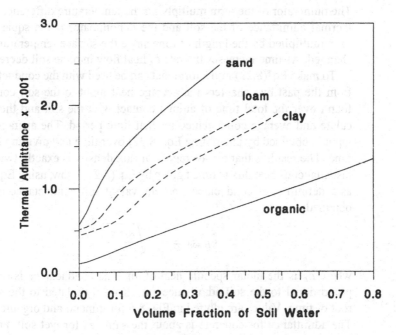

FIGURE 8.5. Thermal admittance of soils in Figs. 8.2 and 8.3.

Mineral soil admittances are also three to five times greater than those for organic soils.

8.5 Heat Transfer from Animals to a Substrate

Equation (8.3) can be solved with a different set of initial and boundary conditions to obtain another result of interest to environmental biophysicists. The practical problem is that of estimating conduction heat loss or heat gain when an animal with body temperature T_b comes in contact with soil or another substrate with initial temperature T_o. The mathematical problem which approximates this is to find temperature as a function of depth and time for a semi-infinite medium of diffusivity κ, and initial temperature T_o when the surface is instantaneously raised to a temperature T_b at time zero. The solution can be found in standard texts on heat transfer. It is:

$$T(z, t) = T_o + (T_b - T_o)\text{erf}\left(\frac{z}{2\sqrt{\kappa t}}\right) \tag{8.21}$$

where erf is the error function, a function which is tabulated in standard mathematical tables. To find the heat flow through the surface of the soil, differentiate Eq. (8.21) with respect to depth to get the temperature gradient, multiply the gradient by the thermal conductivity, and set depth to zero. The result is:

$$G = \sqrt{\frac{k\rho_s c_s}{\pi t}}\,(T_b - T_o) = \frac{\mu_{\text{soil}}(T_b - T_o)}{\sqrt{\pi t}}. \tag{8.22}$$

The numerator of the term multiplying the temperature difference is the thermal admittance of the soil and the denominator is the square root of π multiplied by the length of time since the surface temperature was changed. As time increases the rate of heat flow into the soil decreases.

To make Eq. (8.22) into a form that can be used with the conductances from the past two chapters the average heat input to the soil could be found over the total time of animal contact with the soil and then calculate and average conductance for that time period. The average heat input is obtained by integrating Eq. (8.22) over time and dividing by the time. The result is that the average heat flux density is exactly twice the instantaneous heat flux at time t given by Eq. (8.22). Now, using Eq. (6.8) as a definition of conductance, an equivalent soil conductance can be obtained:

$$g_{H\,\text{soil}} = \frac{2}{c_p}\sqrt{\frac{k\rho_s c_s}{\pi t}} \tag{8.23}$$

where c_p is the molar specific heat of air. The conductance is directly proportional to the soil admittance and inversely related to the square root of time. Values are plotted in Fig. 8.6 for mineral and organic soils. The admittance for concrete is about the same as for wet soil, and the admittance of straw or leaves is similar to that of the organic soil, so

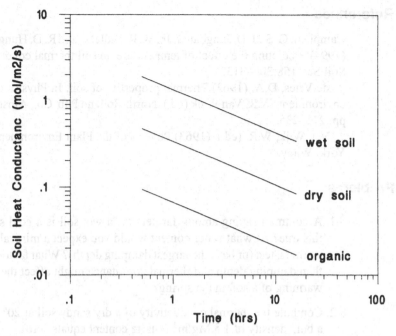

FIGURE 8.6. Thermal conductance of three soil materials averaged for the times shown.

Fig. 8.6 can be used to estimate heat loss or gain for most substrates. The average heat loss decreases by a factor of about five in going from contact periods of a few minutes to contact periods of a day. There is also roughly a factor of five difference between wet soil and dry soil or dry soil and organic material. These numbers should not be completely foreign to your experience. Just compare how you feel about sitting on a concrete bench when the temperature is −20° C to how you feel about sitting on a bale of straw.

Before leaving this subject we need to indicate some cautions and limitations. First, Eqs. (8.21) through (8.23) are for one-dimensional heat flow. For a large animal lying on a substrate for a relatively short time a one-dimensional analysis is probably adequate, but the smaller the animal and the longer the time, the worse the one-dimensional analysis fits the problem. A second point to mention is that the soil conductance is in series with the coat and tissue conductances of the animal. The boundary conditions we used to solve the differential equation are therefore not strictly correct. They should, however, provide a good approximation. The third point is that energy budgets, and therefore conductances, are generally for the whole animal, while these calculations are just for the part of the animal in contact with the solid substrate. The conductance for the whole animal is obtained by multiplying the conductance in Eq. (8.23) by the ratio of area in contact with the substrate to total surface area of the animal.

References

Campbell, G. S., J. D. Jungbauer, Jr., W. R. Bidlake, and R. D. Hungerford. (1994) Predicting the effect of temperature on soil thermal conductivity. Soil Sci. 158:307–313.

de Vries, D.A. (1963) Thermal properties of soil. In Physics of plant environment. W.R. Van Wick (ed.). North Holland Pub. Co., Amsterdam, pp. 210–235.

Van Wijk, W.R. (ed.), (1963) Physics of the Plant Environment, New York: Wiley.

Problems

8.1. A common saying among farmers is "a wet soil is a cold soil." Is this true? At what water content would you expect a mineral soil to warm fastest (or have the largest damping depth)? What factors other than damping depth and thermal admittance might affect the rate of warming of a soil in the spring?

8.2. Compute the thermal conductivity of a dry sandy soil at 20° C with a bulk density of 1.5 Mg/m^3 (quartz content equals zero).

8.3. Rattlesnakes often seek out rocky locations for their dens where they can retreat several meters underground in winter. If the daily near-surface temperature of the rocks is 30° C in summer and −5° C in winter, what is the lowest temperature the snakes would experience during the year if they choose their depth to maximize their temperature?

8.4. Compare the conductive heat losses for a deer on frozen, saturated soil (at 0° C) and a deer on a thick bed of leaves at the same temperature. Assume a body temperature of 37° C, and that 30 percent of the deer's surface is in contact with the substrate for a period of eight hours. Would the conductive losses be about the same, less, or more if the deer were on a frozen, dry soil?

8.5. If the temperature of your bare foot is 35° C and the temperature of the floor is 20° C, calculate the interface temperature between your foot and the floor for a tile floor that has a thermal admittance three times that of your foot, and for a carpet floor that has a thermal admittance 1/3 that of your foot.

Water Flow in Soil 9

The final transport equation that we need to consider is Darcy's law (Eq. (6.4)). This law describes the transport of water in porous materials such as soils. Darcy's law describes most of the water flow that takes place in soils. Since water plays such an important role in the energy balance of soils, plants, and animals, an understanding of at least some simple applications of Darcy's law is important to environmental biophysicists. The processes that are important in determining the water budget of a soil are infiltration of applied water, redistribution of water in the soil profile, evaporation of water from the soil surface, and transpiration of water by plants.

We are mainly interested in applying Darcy's law to problems of one-dimensional water flow, with flow occurring vertically upward or downward. The components of the water potential (Ch. 4) responsible for flow are the matric and gravitational potentials. We can therefore substitute the matric and gravitational potentials for ψ in Eq. (6.4) to obtain:

$$J_w = -K(\psi_m)\frac{d\psi}{dz} = -K(\psi_m)\frac{d\psi_m}{dz} - K\frac{d\psi_g}{dz}$$

$$= -K(\psi_m)\frac{d\psi_m}{dz} - gK(\psi_m). \tag{9.1}$$

Two aspects of this equation make it more complicated mathematically than the equations for diffusion and heat conduction. One is that the hydraulic conductivity has a strong dependence on the dependent variable (matric potential). The other is the flow caused by the gravitational potential gradient. We do not try a frontal attack on Eq. (9.1), but do look for some simple cases for which we can get approximate solutions.

9.1 The Hydraulic Conductivity

The most important factor determining the behavior of Eq. (9.1) is the hydraulic conductivity function. When the soil is saturated with water (all pores filled) the hydraulic conductivity has a value called the saturated conductivity. As the pores drain, the conductivity falls rapidly. With half the pore space drained (roughly field capacity) the conductivity has decreased, typically, by a factor of a thousand. When three-fourths of the pore space has drained (roughly permanent wilting point) the conductivity is only one-millionth of its value at saturation.

A simple equation that gives a good approximation of the hydraulic conductivity is (Campbell, 1985):

$$K(\psi_m) = K_s \left(\frac{\psi_e}{\psi_m} \right)^{2+3/b}$$

or

$$K(\theta) = K_s \left(\frac{\theta}{\theta_s} \right)^{2b+3} \qquad (9.2)$$

where ψ_e is called the air entry water potential and K_s is the saturated conductivity of the soil. The parameter b is the exponent of the moisture release equation, similar to Eq. (4.4):

$$\psi_m = \psi_e \left(\frac{\theta}{\theta_s} \right)^{-b} \qquad (9.3)$$

where θ is the volumetric water content, and θ_s is the saturation water content. The values of b, K_s, and ψ_e depend on soil physical characteristics such as texture.

Saturated conductivity is large for coarse textured soils and small for fine textured, while the inverse is true for ψ_e. Since ψ_e and K_s both depend on the size of the largest pores in the soil, they are not independent, and, in fact are related by the equation: $K_s \psi_e^n = C$, a constant. Table 9.1 gives values of K_s, ψ_e, b and other hydraulic properties for a range of soil texture. This table, with Eqs. (9.2) and (9.3), allows one to estimate the

TABLE 9.1. Hydraulic properties of soils as a function of soil texture (recomputed from Rawls et al. 1992).

Texture	Silt	Clay	$-\psi_e$ J/kg	b	K_s kg s m^{-3}	θ_{-33} m^3/m^3	θ_{-1500} m^3/m^3
sand	0.05	0.03	0.7	1.7	0.0058	0.09	0.03
loamy sand	0.12	0.07	0.9	2.1	0.0017	0.13	0.06
sandy loam	0.25	0.10	1.5	3.1	0.00072	0.21	0.1
loam	0.40	0.18	1.1	4.5	0.00037	0.27	0.12
silt loam	0.65	0.15	2.1	4.7	0.00019	0.33	0.13
sandy clay loam	0.13	0.27	2.8	4	0.00012	0.26	0.15
clay loam	0.34	0.34	2.6	5.2	0.000064	0.32	0.2
silty clay loam	0.58	0.33	3.3	6.6	0.000042	0.37	0.32
sandy clay	0.07	0.40	2.9	6	0.000033	0.34	0.24
silty clay	0.45	0.45	3.4	7.9	0.000025	0.39	0.25
clay	0.20	0.60	3.7	7.6	0.000017	0.4	0.27

hydraulic conductivity and matric potential for various soils at any water content.

9.2 Infiltration of Water into Soil

If water were to pond on a soil surface and the rate at which it infiltrated the soil was measured, results similar to those shown in Fig. 9.1 would be obtained. Results of three experiments are shown, two with vertical infiltration and one with horizontal. Of the two vertical experiments, one was with wet soil and the other dry. For the horizontal column the gravitational gradient is zero, so the matric potential gradient is the only driving force for water flow. The results of the experiments are in agreement with the predictions that would be made using Eq. (9.1). Initial infiltration is dominated by matric forces, so vertical and horizontal infiltration occur at similar rates. The matric potential gradient is smaller for the wet soil than for the dry, so infiltration rate at early times is greater for dry soil than for wet. The water potential gradient for the dry soil is 10000 times greater than that for the wet, yet there is hardly a difference in the infiltration rates. This is because the hydraulic conductivity for the dry soil is much smaller than for the wet.

The influence of matric potential gradients decreases with time, and eventually the gravitational gradient becomes the dominant driving force for flow. Equation (9.1) indicates that the gravitationally-induced flow

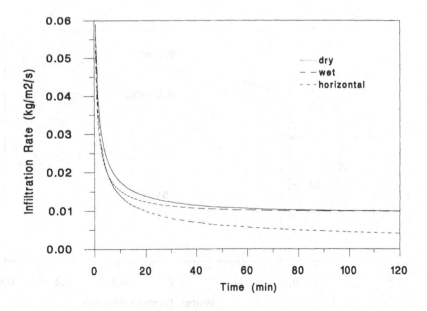

FIGURE 9.1. Vertical infiltration rate for water into initially dry and wet soil, and horizontal infiltration into dry soil.

for a saturated soil is gK_s. The saturated conductivity for the soil in Fig. 9.1 was set at 0.001 kg s m^{-3}, so the final infiltration rate should be 0.0098 kg m^{-2} s^{-1} (0.0098 mm s^{-1}). It can be seen that both curves are approaching this value. The final infiltration rate for the horizontal column is zero.

The important result of the foregoing analysis is that the final infiltration rate can be predicted if the saturated conductivity of soil is known. A simple analysis by Green and Ampt (1911) can be used to estimate the matric-dominated infiltration rate. If we were to measure the water content in the soil as the infiltration shown in Fig. 9.1 occurred, we would obtain the result shown in Fig. 9.2. At each time the soil column consists of essentially wet soil overlying dry soil. A sharp wetting front separates the wet and dry soil. You can see that sharp boundary between the wet and dry soil when you watch water infiltrate dry soil.

The Green–Ampt calculation is made by specifying the location of the wetting front at a point z_f, ignoring the gravitational influence, and approximating the derivative as

$$J_{wi} = -K(\psi_m)\frac{d\psi_m}{dz} = -K_{ave}\frac{\psi_{mf} - \psi_{mi}}{z_f} \tag{9.4}$$

where K_{ave} is the average hydraulic conductivity of the wet soil (called the transmission zone) and ψ_{mf} and ψ_{mi} are the water potentials at the wetting front and the infiltration boundary.

The rate of water storage in the soil is equal to the average change in water content of the transmission zone multiplied by the rate of advance

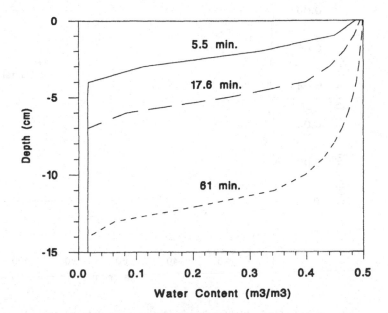

FIGURE 9.2. Water content profiles in soil during infiltration.

of the wetting front. For mass balance, the rate of infiltration must equal
the rate of storage so:

$$K_{ave} \frac{\psi_{mf} - \psi_{mi}}{z_f} = \rho_w \Delta\theta \frac{dz_f}{dt} \qquad (9.5)$$

where ρ_w is the density of water, and $\Delta\theta = (\theta_i + \theta_f)/2 - \theta_o$; θ is the volume fraction of water, and the subscripts i, f, and o are for the infiltration boundary, the wetting front, and the initial water content, respectively. To obtain the position of the wetting front as a function of time separate the variables and integrate:

$$z_f = \sqrt{\frac{2K_{ave}(\psi_{mi} - \psi_{mf})t}{\rho_w \Delta\theta}}. \qquad (9.6)$$

All but t can be expected to be relatively constant during infiltration, so the wetting front will advance linearly with square root of time.

Equation (9.6) can be substutued into Eq. (9.4) to obtain the infiltration rate:

$$J_{wi} = \sqrt{\frac{\rho_w \Delta\theta K_{ave}(\psi_{mi} - \psi_{mf})}{2t}} \qquad (9.7)$$

showing that the infiltration rate is linearly related to the reciprocal of the square root of time. If the data in Fig. 9.1 were replotted with the reciprocal of square root of time as the horizontal axis, the data for the horizontal soil would plot as a straight line. In Ch. 8 we showed that the rate of heat flow into a one-dimensional slab also goes as the inverse square root of time (Eq. (8.22)). It is interesting that the time dependence is the same for heat and water flow, even though the Darcy equation for water is highly nonlinear.

The Green–Ampt approach is strictly only for horizontal infiltration. However, vertical infiltration can be approximated by adding a gravity term to Eq. (9.7). This can then be integrated over time to give an equation for cumulative infiltration:

$$I_w = \sqrt{2\rho_w \Delta\theta K_{ave}(\psi_{mi} - \psi_{mf})t} + gK_{ave}t. \qquad (9.8)$$

The most challenging aspect of using Eq. (9.8) is estimating ψ_{mf}, the matric potential at the wetting front. If we assume that the wetting front is symmetric, then the following approximate expression holds:

$$\psi_{mf} = \frac{2b+3}{b+3} \psi_e \left[1 - \left(\frac{\theta_o}{\theta_s}\right)^{b+3} \right] \approx \frac{2b+3}{b+3} \psi_e \qquad (9.9)$$

where b and ψ_e can be estimated from Table 9.1.

9.3 Redistribution of Water in Soil

When infiltration ceases, water continues to move down into the soil under the influence of matric and gravitational forces. Infiltration was stopped with the final profile shown in Fig. 9.2. The redistribution profiles at four

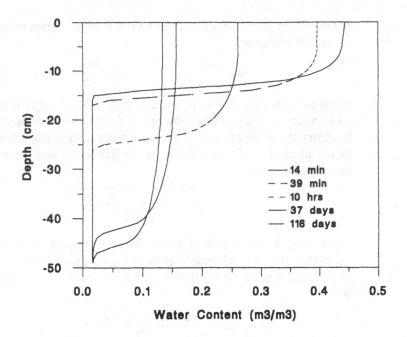

FIGURE 9.3. Redistribution of water in soil following infiltration.

times following are plotted in Fig. 9.3. The water content of the wetted
zone decreases rapidly at first, but the rate of decrease becomes smaller
with time, and eventually there is little change in water content, even
over fairly long times. As the water content of the soil decreases, the
hydraulic conductivity decreases, so the rate of movement of water from
upper to lower layers decreases. This decrease in hydraulic conductivity
with drying allows the soil to store water and gives rise to field capacity.

Darcy's law and some of the characteristics of the redistribution process can be used to get a better understanding of field capacity. Note that
the water content in the wetted part of the profile in Fig. 9.3 is almost
constant with depth, implying that the matric potential is also almost
constant. To do a simple (but rough) analysis of redistribution, the matric
induced flow can be neglected. Then just the gravitational part is left.
Since neither g nor K are ever zero (except when the soil is completely
dry), there will always be some flow out of the wetted zone. We choose
a value ε for the flow out of the wetted portion of the profile, such that
the flow can be considered negligible when compared to water inputs or
other water losses. Using Eqs. (9.1) and (9.2),

$$\varepsilon = gK(\psi_m) = gK_s \left(\frac{\psi_e}{\psi_{fc}} \right)^n \qquad (9.10)$$

where $n = 2 + 3/b$ and ψ_{fc} is the field capacity water potential corresponding to a drainage rate of ε in units of $kg\,m^{-2}s^{-1}$. Substituting the

constant C for the product $K_s \psi_e^n$ and solving for ψ_{fc} gives:

$$\psi_{fc} = -\left(\frac{gC}{\varepsilon}\right)^{1/n}. \tag{9.11}$$

One way to obtain a value for ε is to set it equal to, say, ten percent of evapotranspiration (ET). If ET is 7 mm/day, then ε would be 0.7 mm/day or 8.1×10^{-6} kg m^{-2} s^{-1}. Campbell (1985) gives $C = 10^{-3}$. Putting these values into Eq. (9.11) and assuming $n = 2.5$ gives a field capacity water potential of -17 J/kg. Field capacity determined more empirically is between -10 and -33 J/kg. If ε had been set to one percent of ET, then Eq. (9.11) would give -43 J/kg for ψ_{fc}.

Example 9.1. Use the parameters in Table 9.1 for a silt loam soil to estimate the water content at -10 and -33 J/kg matric potential. Assume $\theta_s = 0.5$ m^3/m^3.

Solution. Solve Eq. (9.2) for θ:

$$\theta = \theta_s \left(\frac{\psi_e}{\psi}\right)^{1/b}.$$

Using this equation, the water contents are

$$\theta_{-33} = 0.5 \left(\frac{-2.1 \text{ J/kg}}{-33 \text{ J/kg}}\right)^{1/4.7} = 0.278 \, \frac{\text{m}^3}{\text{m}^3} \quad \text{and}$$

$$\theta_{-10} = 0.5 \left(\frac{-2.1}{-10}\right)^{1/4.7} = 0.359 \, \frac{\text{m}^3}{\text{m}^3}.$$

The water content at -33 J/kg that is listed in Table 9.1 is 0.33 m^3/m^3. The difference between that value and the one computed here comes from the way the values in Table 9.1 were obtained. The values in the table are averages of many samples for that texture class. Because of the nonlinear nature of Eq. (9.2), the water content from a calculation done using averages of the parameters normally would not be equal to the average of the measured water contents. Another source of uncertainty is the value of θ_s for the soils in Table 9.1.

A number of factors influence field capacity in addition to the ones brought out in this simple analysis. It is actually the hydraulic conductivity function of the soil profile that determines redistribution rates, not the conductivity of a particular location in the profile. If there is significant layering (which there usually is), field capacity will increase. In addition, some texture and density effects appear to cancel in the simple analysis, but may influence field capacity of real profiles. Finally, the matric potential gradient is almost never negligible compared to the gravitational gradient, so there are always matric effects on field capacity. In spite of these limitations, however, several key points are apparent from the analysis. First, we see that field capacity is determined by the ability of the soil to transmit water. Some people suppose that the soil holds water because of the attraction of the matrix for the water. We were able to determine a matric potential at which drainage was considered negligible, but this came from the dependence of conductivity on matric potential, not from the attraction of the matrix for the water. The second point is that a sealed, semi-infinite soil column will continue to drain until it reaches zero water content. There is no point at which drainage ceases. We can therefore think of the soil as a leaky bucket. The size of the leak, however, decreases as the bucket empties.

Some of this is illustrated in Fig. 9.4. The figure shows the water content at a 5 cm depth for the soil in Fig. 9.3. Initially the water content decreases rapidly, but after two to three days the rate of decrease slows. Operationally, field capacity is defined as the water content of the initially wetted soil two to three days after a heavy rain or irrigation when there is no evaporation or transpiration. If Fig. 9.3 were extended to weeks, months, years, or even hundreds of years, water content would continue to decrease. Figure 9.5 extends the graph to about 3 years, and shows that a log-log plot of water content versus time is a straight line.

9.4 Evaporation from the Soil Surface

Water is lost from the root zone of the soil profile in three ways. It can percolate below the root zone, it can evaporate from the soil surface, and it can be taken up by plants. The redistribution calculations just discussed can be used to find the percolation. Evaporation and transpiration still need to be discussed.

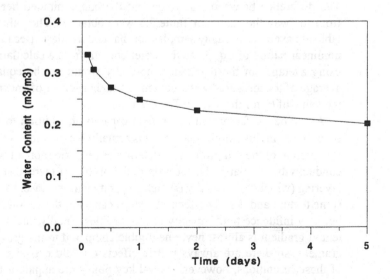

FIGURE 9.4. Water content vs. time at the 5 cm depth in Fig. 9.3.

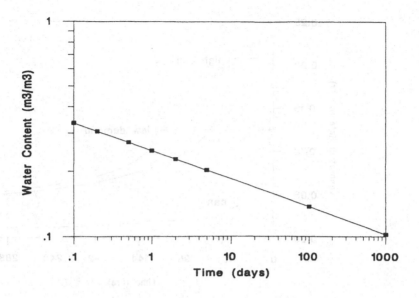

FIGURE 9.5. Redistribution of soil water. Data from Fig. 9.4 plotted on a log-log scale and extended in time.

Figure 9.6 shows the course of evaporation rate over time for three soil drying experiments. Two stages of drying can be identified, a steady constant rate stage and a falling rate stage. The transition between the two occurs when the soil surface becomes dry. The evaporation rate during the first stage is determined by the evaporative demand of the atmosphere. If the demand is high, this stage is short. The lower the evaporative demand, the longer this stage lasts. Coarse textured soils which store little water near the surface have short first stage drying periods. The sand in Fig. 9.6 stores so little water that first stage drying is almost absent.

At the onset of second stage drying, the soil limits the rate of supply to the soil surface. The rate of drying could be determined by calculating the vapor conductance of the dry layer and the vapor pressure difference across it, but the rate is really determined by the ability of the soil to conduct water to the evaporating surface. The form of the solution is similar, again, to the heat flow equation. From the onset of second stage drying the evaporation rate decreases linearly with the inverse of the square root of time, so the cumulative soil surface evaporation during second stage drying (the integral of the rate over time) is proportional to the square root of time:

$$E_{\text{cumII}} = C\sqrt{t - t_1} \qquad (9.12)$$

where t_1 is the time (days) that the first stage drying ends, and C is a constant that depends on soil type. Table 9.2 contains rough estimates of C for several soil textures and also includes approximate values of total cumulative soil surface evaporation for both first and second stage drying.

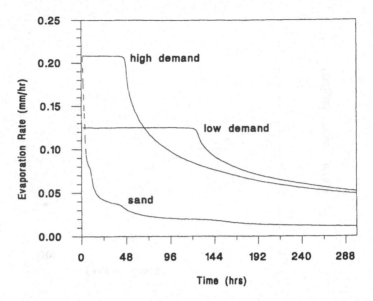

FIGURE 9.6. Evaporation rate for loam at high and low evaporative demand and for sand at high evaporative demand.

Figure 9.7 shows cumulative evaporation versus time for the soils in Fig. 9.6. Interestingly, evaporative demand seems to have little effect on total water evaporated. Early on the high demand soil gets ahead, but the low demand soil stays in first stage evaporation longer and eventually almost catches up. The sand loses much less water than the loam. This is because the surface dries quickly and the coarse material has such a low hydraulic conductivity that it is not able to conduct water to the surface. The ultimate in water conservation is attained with a fine gravel surface, which has almost no storage and very low unsaturated conductivity, but transmits rain downward very readily. The pebble pavement sometimes seen in deserts, where the fine material has been blown away by the wind

TABLE 9.2. Approximate characteristics of soil surface evaporation for several soil textures.

Soil Texture	Cumulative stage 1 evaporation mm	C mm d$^{-1/2}$	Cumulative total evaporation mm
Clay loam	12	5.1	30
Loam	9	4	25
Clay	6	3.5	20
Sand	3	3	10

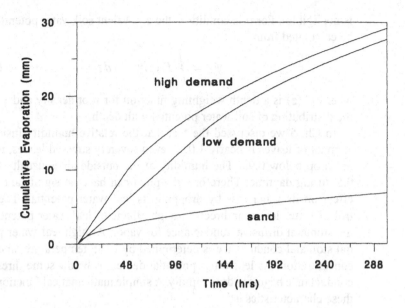

FIGURE 9.7. Cumulative evaporation for the soils in Fig. 9.6.

leaving only coarse material, is a good natural example of a high efficiency
storage system.

9.5 Transpiration and Plant Water Uptake

Liquid water moves from soil to and through roots, through the xylem of
plants, to the leaves, and eventually evaporates in the substomatal cavities
of the leaf. The driving force for this flow is a water potential gradient.
In order for water to flow, the leaf water potential must be below that of
the soil. The entire system is sometimes thought of as being similar to a
resistor network in an electronic circuit where water and current flow are
analogous, and where the potential differences are like voltage differences
in the circuit. Ohm's law is then used to describe the flow of water in the
soil–plant system. The main resistances for liquid water are in the root
and in the leaf, so we can calculate the rate of uptake of water from the
soil as:

$$U = \frac{\psi_s - \psi_L}{R_R + R_L} = \frac{\psi_s - \psi_L}{R_P} \tag{9.13}$$

where ψ_s is the soil water potential, ψ_L is the leaf water potential, and
R_R, R_L, and R_P are the root, leaf, and total plant resistances. The uptake
in Eq. (9.13) should be thought of as uptake per unit area of soil, not per
plant. Campbell (1985) has shown that any distribution of roots and soil
water potential can be represented by a single equivalent potential, which
is the ψ_s in Eq. (9.13). For plants growing in typical field situations, almost
all of the resistance for uptake of water is in the root (the soil resistance

is negligible). For this condition, the equivalent soil water potential can be calculated from

$$\psi_s = \int F_r(z)\psi_s(z)\,dz \qquad (9.14)$$

where $F_r(z)$ is a depth weighting function for root density and $\psi(z)$ is the distribution of soil water potential with depth.

In Ch. 5 we discussed the fact that the relative humidity inside the stomata of leaves is nearly 1.0. In even severely stressed leaves, it does not drop below 0.98. The humidity of the outside air is usually below 0.5 during daytime. Therefore, the plant can have no significant direct effect on its water loss by dropping its leaf water potential. The control of water loss is indirect, through effects of leaf water potential on the stomatal diffusive conductance for vapor. At high leaf water potential stomatal conductance is determined by light, temperature, and CO_2 concentration. As leaf water potential decreases below some threshold, conductance begins to drop rapidly. A simple mathematical function with these characteristics is:

$$\frac{E_p}{E_{pmax}} = \frac{1}{1 + \left(\frac{\psi_L}{\psi_{cl}}\right)^{10}} \qquad (9.15)$$

where E_p and E_{pmax} are the plant transpiration and maximum possible transpiration, and ψ_{cl} sets the threshold leaf water potential for stomatal closure. The power 10 was chosen somewhat arbitrarily. It determines how rapidly the simulated stomata close.

Going back to Eq. (9.13), it can be seen that it describes a linear relationship between uptake rate and leaf water potential (for a given soil water potential). Leaf water potential could decrease indefinitely, and uptake increase indefinitely except for the limit placed on leaf water potential by Eq. (9.15). We are interested in finding what that limit is for any given soil water potential. To do that, we convert Eqs. (9.13) and (9.15) to a dimensionless form. When $\psi_s = 0$ and $U = E_{pmax}$ the leaf water potential will have a value, ψ_{Lm}. Using these values, Eq. (9.13) can be solved for R_P:

$$R_P = \frac{\psi_{Lm}}{E_{pmax}}. \qquad (9.16)$$

Substituting Eq. (9.16) into (9.13), and defining $U^* = U/E_{pmax}$ as a dimensionless uptake rate, $\psi_L^* = \psi_L/\psi_{Lm}$ as a dimensionless leaf water potential, and $\psi_s^* = \psi_s/\psi_{Lm}$ as a dimensionless soil water potential gives

$$U^* = \psi_L^* - \psi_s^*. \qquad (9.17)$$

Equation (9.17) is plotted in Fig. 9.8 for two values of the dimensionless soil water potential (straight lines with positive slope intersecting the horizontal axis at $\psi_L^* = \psi_s^* = 0$ and $\psi_L^* = \psi_s^* = 0.5$ where $U^* = 0$).

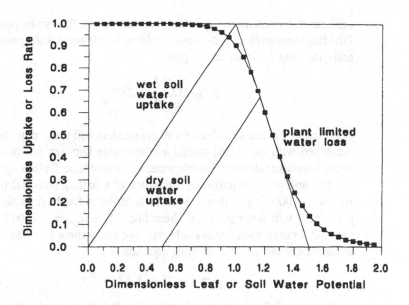

FIGURE 9.8. Dimensionless water uptake and loss.

Equation (9.15) is already in a dimensionless form. The dimensionless transpiration rate can be defined as $E^* = E_p/E_{pmax}$. The ratio of the potentials is the same as the ratio of the dimensionless potentials. Equation (9.15) is also plotted in Fig. 9.8 and it can be seen that the declining part of Eq. (9.15) is closely approximated by a straight line. The equation of the line is

$$E^* = 2(1.5 - \psi_L^*) \qquad (9.18)$$

The maximum or potential uptake rate for a given soil water potential is at the intersection of the uptake and loss lines. Solving Eqs. (9.17) and (9.18) simultaneously to find that point, U_p^*, gives

$$U_p^* = 1 - 2\psi_s^*/3. \qquad (9.19)$$

The actual rate of uptake cannot be higher than this value, but it can be lower if the evaporative demand of the atmosphere is lower. The actual transpiration rate of the plant canopy is therefore equal to the minimum of the evaporative demand of the atmosphere and $E_{pmax}U_p^*$.

These ideas can be related to the depletion of the soil moisture by defining yet another dimensionless quantity, the available water fraction. The available water fraction can be defined as

$$A_w = \frac{\theta - \theta_{pwp}}{\theta_{fc} - \theta_{pwp}} \qquad (9.20)$$

where θ is average water content of the root zone and the subscripts indicate field capacity and permanent wilting water contents. For the simplest case, where it is assumed that water content of the root zone is uniform, substitute from Eq. (9.3) to convert Eq. (9.20) to water potential. Dividing through by ψ_{Lm} to convert to dimensionless soil water potentials, and canceling common terms, gives:

$$A_w = \frac{\psi_s^{-1/b} - \psi_{pwp}^{-1/b}}{\psi_{fc}^{-1/b} - \psi_{pwp}^{-1/b}}. \tag{9.21}$$

Equation (9.21) can be solved for dimensionless soil water potential and combined with Eq. (9.19) to find a relationship between the maximum possible uptake rate and available water in the root zone. Obtaining values for ψ_{fc}^* and ψ_{pwp}^* requires the estimation of a scaling potential ψ_{Lm}. If $\psi_{Lm} = -1000\,\mathrm{J\,kg^{-1}}$, then 1.5 can be substituted for the dimensionless permanent wilt water potential (from Fig. 9.8; $\psi_{pwp} = -1500\,\mathrm{J\,kg^{-1}}$) use 0.02 for the dimensionless field capacity, and assume $b = 5$ to convert to numerical values. The resulting equation is

$$U_p^* = 1 - (1 + 1.3A_w)^{-b}. \tag{9.22}$$

Equation (9.22) is shown plotted in Fig. 9.9. It shows that the potential uptake rate is high until about half of the available water has been extracted. With increasing depletion of soil water the uptake rate falls rapidly. It is important to remember that Fig. 9.9 is not showing the actual uptake rate, it is showing the maximum rate for any given soil water content. If the atmospheric demand is lower than this value, then the uptake will be controlled by the atmospheric demand. The maximum uptake rate when soil is wet is probably about equal to the maximum atmospheric

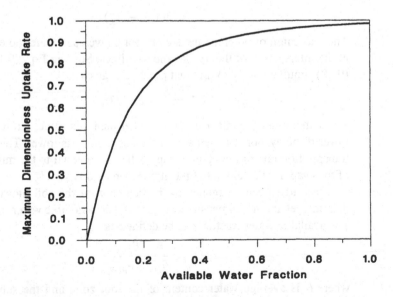

FIGURE 9.9. Maximum rate of plant water uptake as a function of soil available water fraction.

demand on hot days, since plants tend to develop resistances and water potentials which just meet environmental demands.

Example 9.2. Using the same soil properites as in Example 9.1, estimate the available water in mm in a 1 m deep root zone if the permanent wilting water potential is -1500 J kg^{-1}.

Solution.

$$\theta_{pwp} = 0.5\left(\frac{-2.1}{-1500}\right)^{1/4.7} = 0.12\,\frac{m^3}{m^3}.$$

If we assume that field capacity is at -33 J/kg, then, from Example 9.1, $\theta_{fc} = 0.28$ m^3 m^{-3}. The available volumetric water content is $0.28 - 0.12 = 0.16$ m^3 m^{-3}. With a one meter deep root zone, the available water in mm is obtained as 0.16(1000 mm)=160 mm of water. If, on the average, a plant uses 5 mm/day, then a one meter root zone of this soil could potentially provide water for about 32 days.

Example 9.3. If a shallow-rooted plant (rooting depth = 200 mm) is growing in the soil described in Example 9.2 and $E_{pmax} = 5$ mm/day every day, how long will it take before $E_p = 1$ mm/day?

Solution. The total available water in a 200 mm deep profile would be $0.16 \times 200 = 32$ mm. The available water fraction on the first day is $A_w = 1$ so $E_p = 5$ mm/day. It can be assumed that E_p is appropirate for an entire day so that on the second day the available water is 32 mm $-$ 5 mm $= 27$ mm. Thus $A_w = 27/32 = 0.84$ and on the second day the plant uptake rate can be estimated from Eq. (9.22).

$$U_p = E_{p\,max}[1 - (1 + 1.37A_w)^{-b}$$
$$= 5[1 - (1 + 1.37 \times 0.84)^{-4.7}] = 4.9 \text{ mm/day}.$$

On day 3, $A_w = 22.1/32 = 0.69$ so that $U_p = 4.9$ mm/day. This process can be continued until $E_p = 1$ mm/day.

Day	Available water mm	A_w	U_p mm/day
1	32.00	1.00	4.91
2	27.09	0.85	4.87
3	22.22	0.69	4.78
4	17.44	0.54	4.64
5	12.80	0.40	4.36
6	8.44	0.26	3.83
7	4.61	0.14	2.86
8	1.76	0.05	1.44
9	0.31	0.01	0.30

Therefore, 8 days will be required for the transpiration rate to be reduced from 5 to 1 mm/day. Note how quickly the plant runs out of water at the end of 7 days. This is a consequence of the steepness of the curve in Fig. 9.8. Obviously the plant will virtually stop transpiring on the ninth day.

9.6 The Water Balance

We have discussed infiltration, redistribution, evaporation, and transpiration as if they are isolated processes. Of course, they are not, and most go on simultaneously. At any particular time conservation of mass requires that the rate of change in water content in a depth of soil L equal the sum of the inputs and losses. This mass balance equation can be written as

$$\rho_w L \frac{d\theta}{dt} = J_{wi} - J_{wd} - E_s - E_p \qquad (9.23)$$

where the terms on the left represent the rate of storage and the terms on the right are for infiltration (J_{wi}), deep percolation below depth L (J_{wd}), evaporation from the soil surface (E_s), and plant transpiration (E_p). It is not difficult to solve Eq. (9.23) and obtain a record of water content changes in the soil over time, but the solution is only practical using numerical methods on a computer. The methods are covered in detail in Campbell (1985). The analyses presented here are intended to give insight into the processes and how they operate, but do not lead to a full solution of the water balance equation.

References

Campbell, G. S. (1985) Soil Physics with BASIC: Transport Models for Soil-Plant Systems. New York: Elsevier.

Green, W. H. and G. A. Ampt. (1911.) Studies in soil physics: the flow of air and water through soils. J. Agric. Sci. 4:1–24.

Rawls, W.J., L.R. Ahuja, and D.L. Brakensiek (1992) Estimating soil hydraulic properties from soil data. In Indirect Methods for Estimating Hydraulic Properties of Unsaturated Soils. M. th. Van Genucthen, F.J. Leij, and L.J. Lund (eds.) U.C. Riverside Press, Riverside, CA.

Problems

9.1. For a sandy loam soil, estimate the -33 J kg^{-1} and -1500 J kg^{-1} water contents. Assume $\theta_s = 0.45$ m^3 m^{-3}.

a. If the rooting depth is 100 cm, estimate the maximum available water to the plant.

b. If 50 mm of rain falls on this soil in one hour, estimate the runoff if the initial volumetric water content of the soil is 0.1 . Use Eq. (9.8) to estimate the total infiltration in an hour. Compute the water potential at the wetting front from Eq. (9.10), and assume that the average hydraulic conductivity of the transmission zone is the geometric mean of the saturated conductivity and the conductivity at the wetting front (the geometric mean of two numbers is the square root of the product of the numbers). Assume that the potential at the infiltration boundary is zero.

c. If 100 mm of rain falls on this soil in two hours, estimate the runoff if the initial volumetric water content of the soil is $0.1 \ \text{m}^3 \ \text{m}^{-3}$.

d. What is the depth of wetting for each rainfall case above?

e. Estimate the new average water content for the top 100 cm of soil after each of the above rainfalls.

9.2. If $E_{p \, \text{max}}$ is measured to be 5 mm d^{-1} in the sandy loam soil of problem 9.1 immediately after a rain that wets the upper 0.1 m of a root zone to a uniform water content of $0.25 \ \text{m}^3 \ \text{m}^{-3}$, how long will it take before the transpiration decreases to 2.5 mm d^{-1} because of water depletion in the top 0.1 m root zone? Assume that no water is available below the depth of wetting from the rain.

Radiation Basics

10

The modes of energy transport discussed so far (conduction, convection, and latent heat) all are somewhat intuitive. Radiative energy transport, on the other hand, is not intuitive at all. Radiant energy is transferred by photons, discrete bundles of electromagnetic energy that travel at the speed of light ($c = 3 \times 10^{10}$ m/s in vacuum) and behave both as particles and waves. These photons are emitted or absorbed by matter as a result of quantum jumps in electronic energy levels in atoms, or changes in vibrational and rotational energy levels in molecules. The wavelength of the radiation is uniquely related to the photon energy in an equation due to Planck:

$$e = \frac{hc}{\lambda} \tag{10.1}$$

where h is Planck's constant (6.63×10^{-34} J s) and λ is the wavelength of the photon. Thus green photons, having a wavelength of $0.55 \mu m$ would have an energy

$$e = \frac{6.63 \times 10^{-34} \text{ J s} \times 3 \times 10^8 \frac{\text{m}}{\text{s}}}{5.5 \times 10^{-7} \text{ m}} = 3.6 \times 10^{-19} \text{ J.}$$

The energy transferred by a single photon is not generally of interest, but often the energy content of a mole of photons is. This is obtained by multiplying the energy per photon by Avagadro's number (6.023×10^{23}). The energy content of photons at $0.55 \mu m$ wavelength is

$$6.023 \times 10^{23} \frac{\text{photons}}{\text{mol}} \times 3.6 \times 10^{-19} \frac{\text{J}}{\text{photon}} = 2.17 \times 10^5 \frac{\text{J}}{\text{mol}}.$$

This kind of calculation allows conversions between amounts of radiant energy and numbers or moles of photons for a particular wavelength.

The energy of photons could also be expressed as a function of frequency ν of the radiation, since $\nu\lambda = c$, to give $e = h\nu$. Frequency, rather than wavelength is used in some treatments of environmental radiation (Gates, 1980). Advantages of using frequency are a more symmetrical presentation of absorption bands and the ability to show both solar and thermal radiation on a single graph. These advantages are offset somewhat by the loss of detail in the longwave portion of the spectrum and the unfamiliar nature of the units to many biologists. In this presentation we continue to use wavelength.

10.1 The Electromagnetic Spectrum

Photons in natural environments have a wide range of energies and wavelengths. The whole range of photon energies is called the electromagnetic spectrum and is, somewhat arbitrarily, divided into segments or bands according to either the source of the photons or their interaction with living things. Part of the electromagnetic spectrum is shown in Fig. 10.1. The top bar is a logarithmic scale and shows a little more than the full range of wavelengths which are important in radiant energy transport in the natural environment. The upper part of the top bar shows the two important sources of radiation (solar and thermal) and the lower part names three important bands (ultraviolet, visible, and near-infrared). Photons with wavelengths shorter than the ultraviolet end of the spectrum shown are called x-rays and gamma rays. Photons with longer wavelengths are called microwaves and radio waves. The second bar shows two bands of the ultraviolet and the wave bands of the visible colors. The final bar shows some of the biological responses to the different parts of the electromagnetic spectrum.

The previous calculation relating photon numbers and energy content of photons can be extended to cover whole wavebands. If the spectrum of a source is reasonably continuous over a given waveband, the photon flux in that wave band can be calculated from the average energy over the wave band divided by the photon energy at the median wavelength. For example, photosynthetically active radiation (PAR) is the radiation between 400 and 700 nm (Fig. 10.1). For solar radiation at sea level

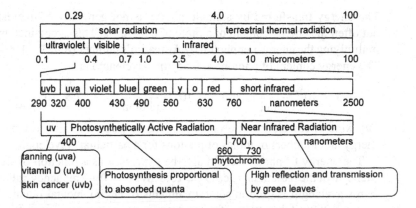

FIGURE 10.1. Part of the electromagnetic spectrum showing names of some of the wavebands and some of the biologically significant interactions with plants and animals.

with a sun zenith angle of 60 degrees, the median wavelength in the 400 to 700 nm waveband is about 550 nm, for which the photon energy is 2.17×10^5 J/mol. This is an approximate but useful number for converting between total energy in the PAR waveband and number of photons for solar radiation. This conversion factor, which is often expressed as 4.6μ mol quanta per J, is within about ten percent of the estimate for cloudy skys and various artificial lights used in growth chambers.

10.2 Blackbody Radiation

As was mentioned, photons are emitted or absorbed because of discrete energy transitions in the emitting or absorbing medium. Each allowable transition produces photons at a single wavelength. If there are many allowable transitions at closely spaced energy levels, the spectral lines tend to merge into an emission or an absorption band. If there are an infinite number of transitions spaced throughout the electromagnetic spectrum, the medium is a perfect radiator or absorber. It will absorb all radiation falling upon it and will radiate the maximum amount of energy that a medium at its temperature is capable of radiating. Such a medium is called a blackbody or full radiator. No such material exists in nature, but some materials approach this behavior over parts of the electromagnetic spectrum. Thus we may speak of a blackbody radiator at visible wavelengths or a blackbody radiator at thermal wavelengths, but would not necessarily expect the same material to be a blackbody in both wave bands. Snow is a very poor absorber of visible radiation, but almost a perfect blackbody in the far infrared.

10.3 Definitions

Radiative transfer results in the transport of energy from a source through a medium to a receiver. This exchange of energy is characterized by the direction of the ray between the source and the receiver, wavelength of radiation, time, coordinates of the point of interest, and the area of the region under consideration. A slight variation on this idea occurs when a surface is being viewed from one direction while an incident ray is from another direction. Although the object being viewed could be considered the source and the viewer could be considered the receiver, it is more useful if both source and view directions are considered simultaneously and so the term bi-directional arises because of the two directions: source and view. Terminology in radiative transfer can be quite complex so the terms used here are only a small subset of the complete terminology. Nicodemus et al. (1977) give a much more extensive set of terms and definitions.

Terms that we use are listed as follows:

Radiant flux (W): The amount of radiant energy emitted, transmitted, or received per unit time.
Radiant flux density (Φ, W/m^2): Radiant flux per unit area.

Irradiance (W/m^2): Radiant flux density incident on a surface.

Radiant spectral flux density (E(λ),W m^{-2} μm^{-1}): Radiant flux density per unit wavelength interval.

Radiant intensity (I, W/sr): Flux emanating from a surface per unit solid angle.

Radiance (N, W m^{-2} sr^{-1}): The radiant flux density emanating from a surface per unit solid angle.

Spectral radiance (W m^{-2} sr^{-1} μm^{-1}) Radiance per unit wavelength interval.

Radiant emittance (W/m^2): Radiant flux density emitted by a surface.

The relationship between various terms describing radiative transfer and energy, time, area, wavelength, and direction, which we previously indicated might be required to specify the radiant energy environment of an organism, can be seen in Fig. 10.2. Obviously, we do not need all of these terms all of the time.

The terms in Fig. 10.2 that are listed on the "hemispherical" side are used extensively in considering balances of energy fluxes on surfaces such as canopies, leaves, lakes, animals, etc. Terms listed on the "directional" side of Fig. 10.2 are used widely in remote sensing, where radiometers (which may be hand-held, aircraft-mounted, or satellite-borne) typically view from a particular direction and sense over a small range of angles about that direction. This small range of angles is typically referred to as the instantaneous field of view (IFOV). For example, an infrared thermometer may be pointed downward at the soil with a zenith angle of 50°

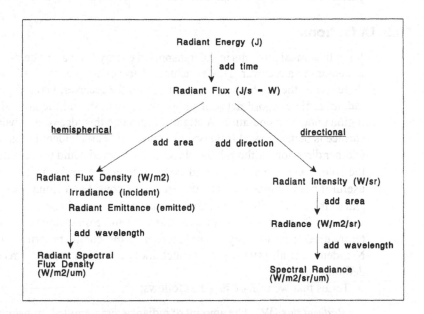

FIGURE 10.2. Relationships among terms for describing the radiant energy environment.

and an azimuth angle of 180° (pointed due South) and its IFOV may be 10°. Thus the measured radiance, which for such an infrared thermometer is probably contained in the wavelength band 8 to 14μm, is appropriate for zenith angles from approximately 45 to 55°, and azimuth angles from 175 to 185°.

Clearly, if enough radiance measurements of a surface are made, integration of these radiance measurements over all the appropriate angles provides an estimate of the radiant flux density. Usually this is not done because of the considerable difficulty associated with making all the directional radiance measurements.

In environmental biophysics we often want to relate remote sensing observations to measurements of radiant or heat fluxes from animals or vegetation. For example, measurements of directional radiometric temperature from an infrared thermometer may provide an estimate of "surface temperature" useful in characterizing the energy budget of a crop. However, great care must be taken in mixing hemispherical and directional quantities, because directional quantities may depend strongly on view angle and have complex relations to hemispherical quantities. In this book, we are mainly concerned with hemispherical quantities.

The most important terms for organism energy balance are the radiant emittance and irradiance. Sets of terms similar to those in Fig. 10.2 are also defined for just the visible portion of the spectrum. When the fluxes are weighted according to the human eye response they are referred to as photometric units, and when they are weighted according to the photosynthetic or quantum response they are referred to as photosynthetic photon flux units. The photometric term corresponding to irradiance is the illuminance, and has units of lumens/m^2 or lux. The photosynthetic term is the photosynthetic photon flux density (PPFD) and has units of mol quanta m^{-2} s^{-1}. As with photosynthesis, one can convert between irradiance and illuminance. The conversion factor depends on the spectral distribution of the radiation, as it does with PAR. For the solar spectrum, 1 mmol m^{-2} s^{-1} $\simeq$ 51 lux. Gates (1980) shows how to do these conversions.

Example 10.1. The irradiance of a surface is 500 W/m^2 in the PAR waveband. What is the PPFD? Assume that the source is solar radiation.

Solution. It was previously determined that energy content of solar radiation in the PAR waveband is 2.35 $\times$ 10^5 J/mol. The PPFD is therefore

$$\text{PPFD} = 500 \frac{J}{m^2 \, s} \times \frac{1 \, mol}{2.35 \times 10^5 \, J} = 2.1 \times 10^{-3} \frac{mol}{m^2 \, s}$$

or 2100μ mol m^{-2} s^{-1}. The irradiance of a horizontal surface under full sun is around 500 W/m^2 in the PAR waveband, so the PPFD just calculated is typical of maximum values measured on clear days.

Example 10.2. An instrument for roughly measuring spectral distribution of radiation in plant environments has blue, red, and near

infrared filters. On the red range it reads out a spectral flux density of
1.3 W m^{-2} nm^{-1}. What is the irradiance in that spectral band?

Solution. Without knowing the bandwidth of the filter, it is impossible
to know the irradiance, but rough estimates can be made. From Fig. 10.1
it can be seen that the red band has a width of 130 nm. If the filter covered
the entire band, then the red irradiance would be 1.3 W m^{-2} nm^{-1} $\times$
130 nm $= 169$ W m^{-2}.

Radiant energy interacts with matter through reflection, transmission,
absorption, and emission. The interaction with the material may depend
on the direction of the incident radiation, the direction from which the
surface is viewed, and the wavelength of the radiation. The wavelength
dependence is recognized in the following definitions.

Absorptivity [$\alpha(\lambda)$]: The fraction of incident radiant flux at a give
wavelength that is absorbed by a material.

Emissivity [$\varepsilon(\lambda)$]: The fraction of blackbody emittance at a given
wavelength emitted by a material.

Reflectivity [$\rho(\lambda)$]: The fraction of incident radiant flux at a given
wavelength reflected by a material.

Transmissivity [$\tau(\lambda)$]: The fraction of incident radiant flux at a given
wavelength transmitted by a material.

Once radiant energy arrives at a receiver it is either absorbed, transmitted, or reflected. Since all of the energy must be partitioned between
these, it can be written that

$$\alpha(\lambda) + \rho(\lambda) + \tau(\lambda) = 1.$$

For a blackbody, $\alpha(\lambda) = 1$, so $\rho(\lambda) = \tau(\lambda) = 0$.

Normally we are interested in the absorption, transmission, reflection,
or emission over an entire waveband, rather than at a single wavelength.
Therefore, for example, a reflection coefficient is defined as

$$\bar{\rho} = \frac{\int_{\lambda_1}^{\lambda_2} \rho(\lambda) E(\lambda)\, d\lambda}{\int_{\lambda_1}^{\lambda_2} E(\lambda)\, d\lambda} \tag{10.2}$$

where $E(\lambda)$ is the radiant spectral flux density of the incident radiation.
Absorption and transmission coefficients are defined similarly. The bands
of interest are generally those shown in Fig. 10.1. We return to this topic
in Ch. 11. We also integrate the emissivity over broad wavebands, and
normalize it as shown in Eq. (10.2), but we do not change its name to
emission coefficient, nor do we usually show the overbar. We just omit the
wavelength in parenthesis. This is because we usually use emissivity to
compute emittance of thermal radiation. The symbol ε therefore always
means the weighted average emissivity over the 4 to 80 μm waveband.

In environmental biophysics we often deal with the thermal radiation exchange between objects and the clear sky. Under these conditions the emissivity in the 8 to 14 μm portion of the 4 to 80 μm wavelength band is most important. On the rare occasion that we are interested in the emissivity of a particular wavelength band, for example the 8 to 14 μm wavelength band of many infrared thermometers, we use $\bar{\varepsilon}_{8-14}$.

From Eq. (10.2) it can be seen that radiative properties of a material (reflection, absorption, transmission coefficients, and emissivity) depend on the wavelength distribution of the source of radiation, as well as the characteristics of the material. A useful way to conceptualize radiation interaction with matter is to always recognize that a source, a medium, and a receiver are involved. The intervening medium is often referred to as a filter. Filters can be natural, such as the atmosphere or water, or they can be artificial and manufactured to accomplish some particular task. Table 10.1 contains several relevant comnbinations to illustrate this.

The following information is always required to assess the interaction between radiation and matter.

1. Radiant flux density (or other measure of radiant energy) as a function of wavelength associated with the source;
2. Transmission or reflection coefficients of the intervening media.
3. Absorptivity as a function of wavelength for the receiver and the response of this receiver as a function of wavelength to the absorbed radiation ($R(\lambda)$).

Consider Case 1 in Table 10.1. Most of the solar radiation is transmitted through the atmosphere (see Fig. 10.5) and is incident on a leaf. The relevant leaf absorptivity depends on the interaction between radiation and the leaf of interest. (For heating of the leaf by direct sun rays, the solar absorptivity is about 0.5 (Fig. 11.5) and all of this absorbed radiation is converted into heat, so the response R(solar) is unity. For photosynthesis, the PAR absorptivity is about 0.85 and photosynthetic rate depends on the magnitude of the absorbed photons.) A response of about 0.03 mol CO_2 fixed in photosynthesis per mol of photons absorbed is typical for corn. Thus the response of the leaf R(PAR) to radiation is 0.03 mol CO_2 mol quanta^{-1}. In Case 2 in Table 10.1, glass and water often are used to filter

TABLE 10.1. Several conbinations of source, intervening medium, and receiver of interest in environmental biophysics.

Case No.	Source	Medium	Receiver
1	Sun	Atmosphere	Leaf
2	Growth Chamber Lamp	Glass + Water	Leaf
3	Sun + Sky	Forest Canopy	Human Eye
4	Soil	Glass window	Infrared Thermometer

out excess heat from lamps in growth chambers so that leaves can absorb high levels of PAR but not be simultaneously exposed to high levels of radiant heat not useful in photosynthesis. This permits better control of leaf temperature. In Case 3, the human eye just happens to have its greatest sensitivity in the green wavelengths (0.55μm) where the absorption of visible radiation by leaves is minimal (See Fig. 11.5); perhaps this is a useful adaptation for survival. Case 4 is interesting because the infrared thermometer has a response of 1.0 in the 8 to 14μm wavelength band ($R(8-14 \mu$m$) = 1.0$). In this wavelength band, thermal radiation emitted by the soil is completely absorbed by the glass because the transmissivity of the glass is zero in this wavelength band. Although the glass transmits 90 percent of the visible radiation "seen" by human eyes, allowing the soil to be seen clearly through the glass, it does not transmit any of the thermal radiation emitted by the soil to the infrared thermometer. Because the glass emits thermal radiation, the infrared thermometer actually measures the temperature of the glass not the soil.

We have shown that emission and absorption of radiation are linked by the same process—that of changing the energy status of the emitting or absorbing atoms or molecules. Thus we would expect the emissivity and absorptivity of a material at a given wavelength to be equal, that is, $\varepsilon(\lambda) = \alpha(\lambda)$, which is a statement of a principle due to Kirchhoff. It is important to recognize that the absorptivity or emissivity values represent only the fraction of *possible* absorption or emission at a particular wavelength, and say nothing about whether or not radiation is actually being absorbed or emitted at that wavelength. For example, carbon black has an emissivity and absorptivity for visible radiation of nearly unity. When carbon black is at room temperature, it may absorb radiation in the solar waveband but emits negligible quantities in that waveband. The radiant emittance at such short wavelengths is near zero not because the emissivity is low, but because there is no energy to be emitted in the solar waveband from such a cold surface. Methods for computing the spectral emittance at a particular wavelength will be given in the next section.

Remote sensing is playing an increasingly important role in plant biophysics. To analyze the interaction of radiation with plant canopies and soil surfaces we need to specifically incorporate directionality into the more general definitions we have just given. The following four definitions are for reflectivity, but corresponding ones could be given for transmissivity.

Bi-directional reflectance (sr^{-1}): The ratio of the reflected radiance from a single view direction to the irradiance from some incident view direction that is confined to a very narrow range of incident angles.

Directional-hemispherical reflectance: The ratio of the reflected radiance integrated over the entire view hemisphere to the irradiance from a single view direction that is confined to a very narrow range of incident angles.

Hemispherical-directional reflectance: The ratio of the reflected radiance from a single view direction to the incident irradiance averaged over the entire incident hemisphere of incoming radiation.

Bi-hemispherical reflectance: The ratio of the reflected radiance integrated over the entire view hemisphere to the incident irradiance averaged over the entire incident hemisphere of incoming radiation.

In remote sensing the term *bi-directional reflectance distribution function* (BRDF) refers to the entire distribution of bi-directional reflectances for all possible view directions. Technically the BRDF is difficult to measure because the solid angle of the source should be infinitesimal for the measurement. The only practical source for BDRF measurements on soils and plant canopies is the sun, and it is not only not an infinitesimal source, but is always accompanied by scattered radiation from the sky. The more useful bi-directional reflectance factor (BRF) is the ratio of the reflected radiance from a single view direction to the reflected radiance from an ideal, perfectly diffuse surface experiencing the same irradiance. Each of the four directional reflectances defined above can have the word factor appended to them if the denominator of the ratio is for an ideal, perfectly diffuse surface. Emission from a surface does not involve incident radiation so the term "directional emissivity" is appropriate.

An interesting fact is worth noting here; the bi-directional reflectance distribution function is not limited to a maximum value of unity like the reflectivity, bi-hemispherical reflectance, and the bi-directional reflectance factors are.

Without dealing with problems of radiation measurement, it may be difficult to understand the applications of these terms. The following example may help. A pyranometer is an instrument sensitive to the solar wavelength band that measures the flux density on a horizontal plane arising from the entire hemisphere above that plane. A pyranometer could be used to measure the irradiance of a crop on a clear day when the solar beam dominates the incident flux. The pyranometer could then be inverted to measure the radiation reflected from the crop. The ratio of reflected to incident radiation would be a bi-hemispherical reflectance, but because the direct beam of the sun dominates the radiance, it would approximate a directional-hemispherical reflectance. The ratio of the same set of measurements made on an overcast day would be a bi-hemispherical reflectance. The two measurements would be similar, but not necessarily equal because of the directional interaction of the surface with the directional solar beam. Instruments that have a narrow view angle, such as aircraft and satellite remote sensing devices, make measurements that approximate the bi-directional reflectance factors. For example, the radiance measured by a narrow IFOV radiometer from some angle divided by the radiant flux density incident on the surface from the sun (irradiance) would approximate the bi-directional reflectance of the surface from the viewing angle. If this narrow IFOV radiometer were used to measure the radiance of a perfectly diffusely reflecting panel with identical orientation and illumination, then the ratio of the measured radiance of the surface and the measured radiance of the perfectly diffusing panel would be the bi-directional reflectance factor.

10.4 The Cosine Law

If a small area is exposed to a point source of radiation, so that the rays of light hitting the surface are nearly parallel, the irradiance of the surface depends on its orientation with respect to the radiant beam. This is easily seen by considering the area on a surface covered by a beam of parallel light of fixed size as its angle with respect to a normal to the surface increases (Fig. 10.3). The radiant flux density on the area perpendicular to the direction of the beam remains constant, but the beam covers a larger and larger area as the zenith angle (θ) increases, so the flux density at the surface decreases. If the area covered by the beam at normal incidence is A_p and the area at angle θ is A then $A_p/A = \cos \theta$. This leads directly to Lambert's cosine law:

$$\Phi = \Phi_o \cos \theta \tag{10.3}$$

where Φ_o is the flux density normal to the beam, Φ is the flux density at the surface, and θ is the angle between the radiant beam and a normal to the surface, which is referred to as the zenith angle.

The only common source of parallel light in natural environments is the sun, and Lambert's law is used to calculate the direct solar irradiance of slopes, walls, leaves, or animals. To do the calculation, Φ_o and the

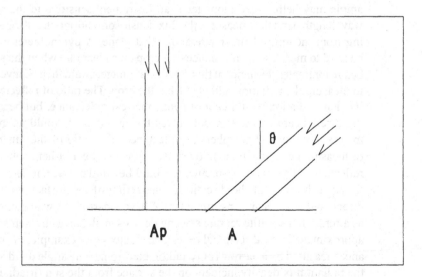

FIGURE 10.3. The area covered by a beam of parallel light increases as the angle θ between the beam and a normal to the surface increases.

angle the sun makes with a normal to the surface need to be known. Equation (10.3) can also be used to find the irradiance of a surface when the radiance of the surroundings is known, as shown in the following example.

Example 10.3. A unit area on the ground is illuminated by a hemisphere of isotropic radiation with a radiance of N W m^{-2} sr^{-1} (isotropic means that the radiance is constant for all incident directions). What is the irradiance of the surface?

Solution. The irradiance of the surface by a small increment of solid angle $d\Omega$, which makes an angle θ with a normal to the surface, is $N \cos \theta d\Omega$. The product of the radiance and the solid angle gives the flux density of radiation on a surface perpendicular to the direction of that radiation. The cosine of the angle converts this to flux density on the horizontal ground. To find the total irradiance of the surface, integrate the radiance of the hemisphere over all solid angles that are visible from the surface. If ψ is the azimuth angle, then $d\Omega = \sin \theta \, d\theta \psi$. The irradiance is therefore

$$\Phi = \int_o^{2\pi} \int_o^{\pi/2} N \sin \theta \cos \theta \, d\theta \, d\psi = \pi N (\text{W/m}^2)$$

so the irradiance of a surface under isotropic radiation is always π times the radiance. Here N is constant and can be taken out of the integral. If the radiation were not isotropic the irradiance of the surface could be found in the same way, but the angular distribution of N would need to be part of the integration.

An ideal diffusely reflecting surface, sometimes referred to as an ideal Lambertian surface, has a radiance that is proportional to the cosine of the angle between a normal to the surface and the view direction. The directional-hemispherical reflectance of such a surface is unity. In remote sensing, various surfaces are used to approximate a Lambertian surface, such as molded Halon or barium sulfate. The reflection coefficients of natural surfaces, such as lakes, vegetation, soils, and rocks may differ substantially from that of an ideal surface. Therefore, care must be taken when specifing what is meant when referring to the "reflectance" of some natural surface. Biophysicists and micrometeorologists have used the term albedo to refer to the bi-hemispherical reflectance integrated over the entire solar spectrum. This operational approach gets around some of the complexity.

10.5 Attenuation of Radiation

Parallel monochromatic radiation propagating through a homogeneous medium that attenuates the beam will show a decrease in flux density described by Bouguer's or Beer's law:

$$\Phi = \Phi_o \exp(-kz) \tag{10.4}$$

where Φ_o is the unattenuated flux density, z is the distance the beam travels in the medium, and k is an extinction coefficient (m^{-1}) for the medium. We use this law to describe light penetration in the atmosphere and in crop canopies. The law strictly applies only for wavebands narrow enough that k remains relatively constant over the waveband. It is often used, however, for much broader bands of radiation. We use it to describe the attenuation of the entire solar spectrum in the atmosphere. For broad wavebands attenuation may not exactly follow an exponential law. If changes in kz are not too great, however, Eq. (10.4) can still give a good approximation of the attenuation of broadband radiation with distance.

Example 10.4. When Eq. (10.4) is used to find the attenuation of solar radiation by the atmosphere, distance is measured in terms of an airmass number, $m = 1/\cos\theta$. The extinction coefficient is then per airmass, rather than per meter. If measured solar beam radiation were 871, 785, and 620 W/m^2, at zenith angles (the angle between the solar beam and a vertical) of 30, 45, and 60 degrees, find the extinction coefficient, k, and Φ_o.

Solution. Taking the logarithm of both sides of Eq. (10.4) gives $\ln\Phi = \ln\Phi_o - km$, so a regression of $\ln\Phi$ on m will have a slope of $-k$ and an

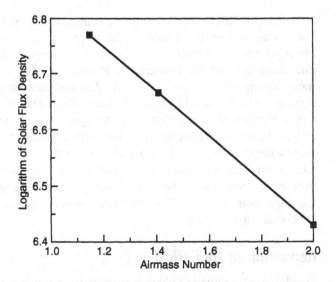

FIGURE FOR EXAMPLE 10.4.

intercept of $\ln\Phi_o$. The necessary computations are in the table. A graph of the data is shown in the accompanying figure.

Zenith angle θ	Airmass m	Φ	$\ln \Phi$
30	1.15	871	6.770
45	1.41	785	6.666
60	2.00	620	6.430

The slope is -0.4 so $k = 0.4$ per airmass. The intercept is 7.23 so $\Phi_o = \exp(7.23) = 1380$ W/m^2; this value is near the solar constant of about 1360 W/m^2.

10.6 Spectral Distribution of Blackbody Radiation

One of the major breakthroughs of modern physics was the discovery of a correct model for the spectral distribution of blackbody radiation. Classical approaches predicted that the amount of energy emitted by a surface would increase without bound as the wavelength of the radiation decreased. This implied that all of the energy in the universe would ultimately be funneled to short wavelengths and emitted; a situation referred to as the "ultraviolet catastrophe." The catastrophe was the fault of the model, of course, not of nature. This was all solved by Planck's quantum hypothesis, that energy is emitted in discrete packages, or quanta, whose energy and wavelength are related by Eq. (10.1). Planck's model for the radiant spectral flux density from a blackbody radiator is

$$E_b(\lambda, \mathbf{T}) = \frac{2\pi hc^2}{\lambda^5[\exp(hc/k\lambda\mathbf{T}) - 1]}. \tag{10.5}$$

Here $E_b(\lambda, \mathbf{T})$ (W/m^3) is the radiant spectral flux density or spectral emittance, $\mathbf{T}$ is the kelvin temperature, h is Planck's constant, and k is the Boltzmann constant (1.38×10^{-23} J/K). Blackbody spectra are plotted in Fig. 10.4 for sources at 6000 K and 288 K, corresponding roughly to sun and earth emittance spectra. Note that we have used a logarithmic scale for wavelength so that both spectra can be shown in the same graph. The two spectra overlap slightly between 3 and 4 μm, but the amount of energy in the overlap is negligible. We therefore specify 4 μm as the top end of the solar spectrum and the bottom end of the terrestrial thermal spectrum. The scale for the sun emittance is 10^6 larger than for the earth. Essentially all of the energy emitted by the earth comes from the sun, but the earth intercepts only a very small fraction of the energy the sun emits.

The wavelength of peak spectral emittance is a function of temperature of the emitting surface, as can be seen from Fig. 10.4. The wavelength at peak emittance (on a wavelength basis) is found by differentiating

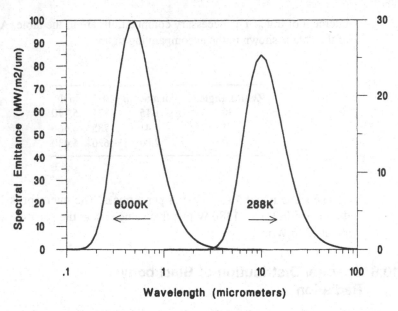

FIGURE 10.4. Emittance spectra for 6000 K and 288 K blackbody sources approximating emission from the sun and the earth.

Eq. (10.5) with respect to λ, setting the result to zero, and solving for temperature. The result is called Wien's law:

$$\lambda_m = \frac{2897\,\mu mK}{T} \tag{10.6}$$

where λ_m is in micrometers. The wavelengths for peak emittance for the 6000 K and 288 K sources in Fig. 10.4 are 0.48 and 10 μm respectively.

10.7 Spectral Distribution of Solar and Thermal Radiation

The actual extraterrestrial solar spectrum is shown in Fig. 10.5. Here it is plotted using a linear wavelength scale. It has nearly the same shape as a blackbody source at 6000 K, and the wavelength at peak emission is between blue and green, as expected from the Wien's law calculation. As solar radiation passes through the atmosphere of the earth, some wavelengths are almost completely absorbed. The ozone layer in the stratosphere absorbs much of the ultraviolet radiation. Water vapor is the main absorber in the infrared.

The strong absorption of short wavelength ultraviolet radiation by ozone is of particular importance to living organisms. Referring to Fig. 10.1, it can be seen that radiation at these wavelengths can cause skin cancer. It is actually capable of inducing mutation in any genetic material and of germicidal action. This is the reason for the recent concern over destruction of the ozone layer by release into the atmosphere of chlo-

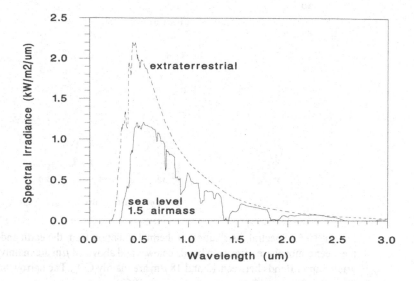

FIGURE 10.5. Spectral irradiance of the sun just outside the atmosphere and at sea level through a 1.5 airmass atmospheric path. Atmospheric absorption at short wavelengths is mainly from ozone. At long wavelengths it is mainly from water vapor (redrawn from Gates, 1980).

rofluorocarbons. These compounds destroy ozone and could increase the flux of harmful ultraviolet radiation at the surface of the earth.

Energy over the entire solar spectrum is reduced by Rayleigh (small-particle) and Mie (large-particle) scattering. Rayleigh scattering is from the molecules of air and is most pronounced at short wavelengths so the scattered radiation is blue. This is the source of the blue color of the sky. Blue wavelengths are preferentially scattered out of the solar beam, causing the sun to appear red. Mie scattering is from dust, smoke, and other aerosols in the atmosphere. Conditions can exist which result in preferential scattering of long wavelengths by Mie scatterers, but generally there is little wavelength dependence.

About half of the energy in the solar spectrum is at wavelengths shorter than 0.7 μm and half at longer wavelengths (actually about 45 percent is in the visible and 55 percent in the near-infrared). The spectrum changes with solar zenith angle, cloudiness, and atmospheric composition, but the distribution between visible and infrared remains almost unchanged. Many of our computations require that the energy content of these two wavebands are known. Nature made it easy for us by consistently partitioning approximately half to each.

The mean emission of the earth approximates that of a blackbody with a temperature of 288 K. The spectral emittance for such a blackbody is shown in Fig. 10.6. Almost all of the radiation is at wavelengths longer than 4 μm and the wavelength at peak emission is 10 μm. The emittance

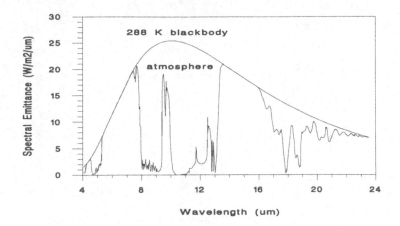

Wavelength (um)

FIGURE 10.6. Spectral distribution of thermal radiation from the earth and from the clear atmosphere. Emission bands below 8 and above 18 μm are mainly from water vapor. Bands between 13 and 18 μm are mainly CO_2. The narrow band at 9.5 μm is from ozone (redrawn from Gates, 1962).

spectrum of most terrestrial objects is similar to Fig. 10.6, but the peak location and height shift somewhat depending on the surface temperature.

Thermal radiation is emitted and absorbed in a clear atmosphere mainly by water vapor and CO_2, with a narrow ozone absorption band around 9.5 μm. Infrared radiation is absorbed or emitted as a result of changes in the vibrational and rotational energy levels of molecules. Water vapor, CO_2, and O_3 are the only common atmospheric constituents with energy levels that are excited by thermal radiation. An atmospheric emittance spectrum is shown in Fig. 10.6 along with the 288 K blackbody spectrum. It can be seen that the atmosphere acts almost like a blackbody in some wavebands where there is strong absorption and emission. In other wavebands the absorptivity and emissivity are low. The "window" between 8 and 13 μm has particular importance. This coincides with the blackbody emission peak for the earth at 288 K. Much of the radiation emitted by the earth in these wavelengths is not absorbed by the atmosphere and is lost to space.

10.8 Radiant Emittance

The total radiant energy emitted by a unit area of surface of a blackbody radiator is found by integrating Eq. (10.5) over all wavelengths. The result is the Stefan–Boltzmann law:

$$B = \sigma T^4 \tag{10.7}$$

where B is the emitted flux density (W/m^2), $\mathbf{T}$ is the Kelvin temperature, and σ is the Stefan–Boltzmann constant (5.67×10^{-8} W m^{-2} K^{-4}). Values of B at various temperatures are given in Table A.3.

Example 10.5. Find the average radiant emittance of the earth and the sun.

Solution. The earth approximates a blackbody radiator emitting at 288 K. The average emittance of the earth is therefore (5.67×10^{-8} W m^{-2} K^{-4} $\times$ (288 K)^4 = 390 W/m^2. The sun emittance is approximately that of a blackbody at 6000 K. Using Eq. (10.7) again, the energy emitted is therefore 73 MW/m^2 at the surface of the sun.

The energy emitted by nonblackbodies is given by:

$$\Phi = \int \varepsilon(\lambda) E_b(\lambda, t) \, d\lambda \qquad (10.8)$$

where $\varepsilon(\lambda)$ is the spectral distribution of emissivity and $E_b(\lambda, T)$ is from Eq. (10.5). A gray body is one which has no wavelength dependence of the emissivity so the integration produces:

$$\Phi = \varepsilon B. \qquad (10.9)$$

Natural surfaces are not perfect gray bodies, and the result of applying Eq. (10.8), is a power of **T** that is not exactly 4 in the Stefan–Boltzmann equation. In practice, for the range of normal terrestrial temperatures, all bodies can be treated as gray bodies and Eqs. (10.7) and (10.9) can be used with an appropriate average emissivity. This approach also works for computing atmospheric emittance, even though the atmosphere is far from a gray body. In the next chapter we show that the emissivity of most natural surfaces is between 0.95 and 1.0. For most of our calculations we assume a value of 0.97. The emissivity of a clear atmosphere, however, is much lower, as can be seen in Fig. 10.6. Clouds increase the emissivity of the atmosphere and the emissivity of a completely overcast sky with a low cloud base is near unity.

Several empirical formulae are available for computing estimates of clear sky emissivity. One with reasonable theoretical justification is (Brutsaert, 1984):

$$\varepsilon_{ac} = 1.72 \left(\frac{e_a}{\mathbf{T}_a} \right)^{1/7} \qquad (10.10)$$

where e_a is the vapor pressure (kPa) measured at height of one to two meters and $\mathbf{T}_a$ is the air temperature (kelvins). The reasoning behind this formula is that atmospheric thermal radiation is primarily a function of the water vapor concentration in the first few kilometers of the atmosphere and is most strongly dependent on the vapor concentration in the first few hundred meters. Thus a measurement of vapor concentration at a height of one to two meters, combined with estimates of vapor and temperature profiles to 5 km, can be used to estimate emissivity.

Since vapor pressure and minimum temperature are strongly correlated, correlations have also been made between temperature at a height of one to two meters and clear sky emissivity. Swinbank (1963) suggests the formula:

$$\varepsilon_{ac} = 9.2 \times 10^{-6} T_a^2. \tag{10.11}$$

Brutsaert (1984) reconciled Eqs. (10.10) and (10.11) using an empirical correlation of temperature and vapor pressure. We use Eq. (10.11) for most of our computations, and values are listed in Table A.3. If vapor pressure data are available, Eq. (10.10) is probably preferable.

Clouds have an emissivity of one, so when clouds are present, atmospheric emissivity is higher than for a clear sky. The atmospheric emittance on cloudy days can be estimated by adding the energy emitted by the clear portions of the sky to the energy emitted by the clouds. Monteith and Unsworth (1990) give the simple relationship

$$\varepsilon_a(c) = (1 - 0.84c)\varepsilon_{ac} + 0.84c \tag{10.12}$$

where c is the fraction of the sky covered by cloud and ε_{ac} is given by Eq. (10.11) or (10.10). When c is zero, $\varepsilon_a(c) = \varepsilon_{ac}$. When $c = 1$, $\varepsilon_a(c) = 0.84 + 0.16\varepsilon_{ac}$. At a temperature of 20° C, this gives a sky emissivity of 0.97.

Example 10.6. Compare clear sky and completely overcast sky emittance when air temperature is 20° C.

Solution. Using Eq. (10.7) or Table A.3 the black body emittance can be found. Equation (10.7) gives

$$B = 5.67 \times 10^{-8} \times (20 + 273.16)^4 = 419 \text{ W m}^{-2}$$

The clear sky emissivity (Eq. (10.11) or Table A.3) is $9.2 \times 10^{-6} \times (273.16 + 20)^2 = 0.79$. The emissivity for a completely overcast sky (Eq. (10.12), $c = 1$) is $0.16 \times 0.79 + 0.84 = 0.97$. The emittances are

clear sky: $0.79 \times 419 = 331 \text{ W m}^{-2}$

cloudy sky: $0.97 \times 419 = 406 \text{ W m}^{-2}$.

The cooling and frost that occur on clear nights are sometimes explained as "radiation being lost to outer space." This description is both overly dramatic and wrong. The difference between a clear night and a cloudy night is not the outgoing but the incoming radiation. The ground receives less radiation from the atmosphere on clear nights (and days) than on cloudy ones.

We could compute the emittance of the sun just as we do the earth and atmosphere, but this has little value for environmental biophysics. We assume the output of the sun is constant and use that constant, measured value for all of our calculations. The mean radiant flux density outside the atmosphere of the earth and normal to the solar beam is about 1360 W m^{-2}. This value is known as the solar constant. The actual flux density varies by about ±1.5 percent due to random variations in solar activity and ±3.5 percent annually due to the predictable variation in earth-sun distance. These variations are much smaller than other sources of uncertainty in the radiation budget, so we do not try to account for them in our calculations, and assume that the solar input just outside the atmosphere is constant and equal to the solar constant.

References

Brutsaert, W. (1975) On a derivable formula for long-wave radiation from clear skies. Water Resour. Res. 11: 742–744.

Brutsaert, W. (1984) Evaporation into the Atmosphere: Theory, History, and Applications. Boston: D. Reidel.

Gates, D. M. (1980) Biophysical Ecology. New York: Springer Verlag.

Monteith, J. L. and M. H. Unsworth (1990) Principles of environmental Physics (2nd ed.). London: Edward Arnold.

Nicodemus, F. E., J. C. Richmond, J. J. Hsia, I. W. Ginsberg, and T. Limperis (1977) "Geometrical considerations and nomenclature for reflectance." NBS Monograph 160, U.S. Dept. Commerce/National Bureau Standards, 1977.

Problems

10.1. The median wavelength for solar radiation in the 0.3 to 3 μm wave band is approximately 0.7 μm. If the irradiance is 1 kW/m^2, what is the photon flux?

10.2. When the energy flux density in the 400 to 700 nm waveband is 200 W/m^2, what is the photon flux density of PAR (mol m^{-2} s^{-1})?

10.3. When photon flux density is 1000 μmol m^{-2} s^{-1} in the 400 to 700 nm waveband, what is the energy flux density (W/m^2) in the PAR waveband? What is the flux density of solar radiation (all wavelengths)?

10.4. What is the wavelength of peak emittance for a 2800 K incandescent light bulb?

10.5. If your mean surface temperature is 28° C, what is your emittance? If the mean wall temperature in the room in which you are standing is 20° C, what is the average thermal irradiance at your surface? Estimate your net radiant heat loss. Assume $\varepsilon_{surf} = 0.97$ and $\varepsilon_{wall} = 1.0$.

10.6. Compare the radiant emittance of a clear sky and a completely overcast sky, both at 0° C. How much additional radiant energy does the ground receive on an overcast night?

Radiation Fluxes in Natural Environments

11

Before beginning a detailed discussion of radiant energy budgets of plants, animals, canopies, and soils, we need to determine what information will be required and how to obtain that information. An environmental biophysicist may approach the study of radiant energy exchange in two different ways. For the first, detailed observations of radiant flux densities to and from an organism are needed to compute a detailed energy budget. These detailed observations must be obtained by direct measurement at the time the energy budget is being determined. The second type of study simulates the behavior of parts of an ecosystem. Knowing the exact value of a radiant flux density may not be as important as having the correct relationship among variables. Models of the fluxes, extended from the basics covered in Ch. 10, are used for studies of the second type. The models can be counted on to give reasonable estimates ($\pm 10\%$) of average flux densities, but they are usually not adequate as substitutes for careful field measurements of radiant fluxes for detailed energy budget studies. This chapter presents models for estimating solar and thermal radiant fluxes in the natural environment.

Streams of solar radiation received by an organism are

1. beam, or direct radiation, directly from the sun,
2. diffuse radiation, scattered by sky and clouds, and
3. reflected radiation from terrestrial objects.

Separating solar radiation into at least this many components is necessary because the amounts and directional characteristics are different for each. Beam radiation is highly directional and irradiance at a surface is determined using Lambert's cosine law (Eq. (10.3)). Diffuse sky and reflected ground radiation are scattered from all directions. The diffuse irradiance at a surface is computed by integrating the radiance of the surroundings using the procedure given in Ch. 10. To find the components of the solar radiation budget the following need to be known: the flux density of solar radiation perpendicular to the solar beam, the angle the beam makes with the absorbing surface, and the flux densities of sky diffuse and ground reflected radiation.

The thermal radiation in outdoor environments comes from two distinct sources, the ground and the sky. In enclosures or under vegetation the thermal radiation comes from the walls of the enclosure or from the canopy. All of this radiation is diffuse, and the irradiance, again, is computed by integrating the radiance over all sources surrounding the organism surface. Chapter 10 shows how to compute the radiance of these sources.

11.1 Sun Angles and Daylength

The location of the sun in the sky is described in terms of its altitude (β, elevation angle above the horizon) or zenith angle (ψ, the angle measured from the vertical) and its azimuth angle (AZ, angle from true north or south measured in the horizontal plane). Several coordinate systems are possible with azimuth angles:

1. mathematical: zero degrees is south and angles increase in the counter clockwise direction from 0 to 360°,
2. compass: zero degrees is north and angles increase from 0 to 360° in a clockwise direction, and
3. astronomical: zero degrees is south and positive angles increase from 0 to 180° in a clockwise direction; the counter clockwise direction is labeled with negative angles from 0 to −180°.

We use the mathematical coordinate representation. Elevation and zenith angles are related by $\beta = 90 - \psi$ (angles in degrees). The zenith angle of the sun depends on the time of day, the latitude of the site, and the time of year. It is calculated from

$$\cos \psi = \sin \beta = \sin \phi \sin \delta + \cos \phi \cos \delta \cos[15(t - t_0)] \quad (11.1)$$

where ϕ is the latitude, δ is solar declination, t is time, and t_0 is the time of solar noon. The earth turns at a rate of 360° per 24 hours, so the 15 factor converts hours to degrees. Time, t is in hours (standard local time), ranging from 0 to 24. Latitude of a site is found in an atlas. Solar declination ranges from +23.45° at summer solstice to −23.45° at winter solstice. It can be calculated from

$$\sin \delta = 0.39785 \sin[278.97 + 0.9856J + 1.9165 \sin(356.6 + 0.9856J)] \quad (11.2)$$

where J is the calendar day with $J = 1$ at January 1. Some values of δ are given in Table 11.1.

The time of solar noon is calculated from

$$t_o = 12 - LC - ET \quad (11.3)$$

where LC is the longitude correction and ET is the equation of time. LC is +4 minutes, or +1/15 hour for each degree you are east of the standard meridian and −1/15 hour for each degree west of the standard meridian. Standard meridians are at 0, 15, 30, ..., 345°. Generally time zones run approximately +7.5 to −7.5° either side of a standard meridian,

TABLE 11.1.   Solar declination and Equation of Time for various dates.

Date	Day	Declin. Degree	E. T. hours	Date	Day	Declin. Degree	E. T. hours
Jan 1	1	−23.09	−0.057	Jun 29	180	23.26	−0.055
Jan 10	10	−22.12	−0.123	Jul 9	190	22.46	−0.085
Jan 20	20	−20.34	−0.182	Jul 19	200	20.97	−0.103
Jan 30	30	−17.88	−0.222	Jul 29	210	18.96	−0.107
Feb 9	40	−14.95	−0.238	Aug 8	220	16.39	−0.094
Feb 19	50	−11.57	−0.232	Aug 18	230	13.35	−0.065
Mar 1	60	−7.91	−0.208	Aug 28	240	9.97	−0.022
Mar 11	70	−4.07	−0.170	Sep 7	250	6.36	0.031
Mar 21	80	−0.11	−0.122	Sep 17	260	2.58	0.089
Mar 31	90	3.84	−0.072	Sep 27	270	−1.32	0.147
Apr 10	100	7.62	−0.024	Oct 7	280	−5.21	0.201
Apr 20	110	11.23	0.017	Oct 17	290	−9.00	0.243
Apr 30	120	14.50	0.046	Oct 27	300	−12.55	0.268
May 10	130	17.42	0.060	Nov 6	310	−15.76	0.273
May 20	140	19.82	0.059	Nov 16	320	−18.56	0.255
May 30	150	21.66	0.043	Nov 26	330	−20.80	0.213
Jun 9	160	22.86	0.015	Dec 6	340	−22.40	0.151
Jun 19	170	23.43	−0.019	Dec 16	350	−23.26	0.075
				Dec 26	360	−23.38	−0.007

but this sometimes varies depending on political boundaries. An atlas can be checked to get both the standard meridian and the longitude. The equation of time is a 15 to 20 minute correction which depends on calendar day. It can be calculated from

$$ET = \frac{-104.7 \sin f + 596.2 \sin 2f + 4.3 \sin 3f - 12.7 \sin 4f - 429.3 \cos f - 2.0 \cos 2f + 19.3 \cos 3f}{3600}$$

(11.4)

where $f = 279.575 + 0.9856J$, in degrees. Some values for ET are also given in Table 11.1.

The azimuth angle of the sun can be calculated from

$$\cos AZ = \frac{-(\sin \delta - \cos \psi \sin \phi)}{\cos \phi \sin \psi}$$

(11.5)

where ψ is the zenith angle, calculated from Eq. (11.1); and AZ is in degrees, measured with respect to due south, increasing in the counter clockwise direction so 90° is east. Afternoon azimuth angles can be calculated by taking 360° minus the AZ calculated from Eq. (11.5), or by multiplying the result of Eq. (11.5) by −1; these two cases being only two of many possibilities for labeling the azimuth. Using Eqs. (11.1) and (11.5) the sun paths can be plotted for different latitudes, times of the year, and times of the day. Figure 11.1 shows some examples for latitudes of 0, 25, 50, and 75°. Note that the angle scale in Fig. 11.1 is different from that calculated from Eq. (11.5). Several azimuth-scale

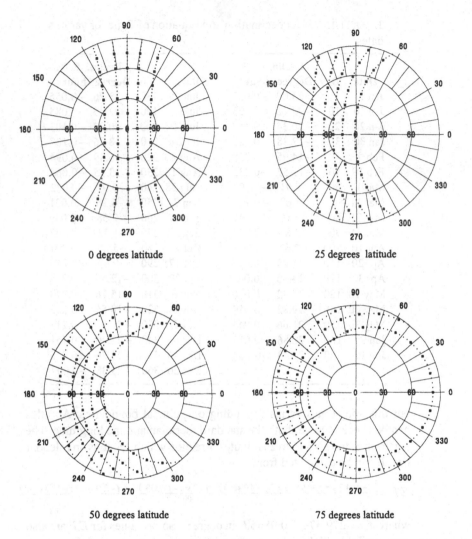

0 degrees latitude

25 degrees latitude

50 degrees latitude

75 degrees latitude

FIGURE 11-1. Sun tracks at declination angles of −23.5, −10, 0, 10, and 23.5° for four different latitudes. Zenith angle grids are the concentric circles. Azimuth angles are shown around the outer circle. North is 0°, east is 90°. Large dots are at one hour time increments.

labeling conventions exist, such as mathematical, astronomical or geographical, so that the reader should be prepared to convert between various labeling conventions.

Equation (11.1) can be rearranged to solve for daylength. It works best to write the equation in terms of the half daylength, h_s, which is the time (in degrees) from sunrise to solar noon. The half daylength is

$$h_s = \cos^{-1}\left(\frac{\cos\psi - \sin\phi\sin\delta}{\cos\phi\cos\delta}\right) \qquad (11.6)$$

where $\cos\psi = 0$ for a geometric sunset (no atmospheric refraction) and a small negative value when twilight is considered. The time of sunrise

(t_r) is the time of solar noon (t_o) minus the half daylength divided by 15 (to convert degrees to hours):

$$t_r = t_o - h_s/15 \tag{11.7}$$

and daylength in hours is twice the half daylength in degrees divided by $15°/hr$ $(2h_s/15)$. Therefore, a 12-hour daylength corresponds to $h_s = 90°$.

To find the time of actual sunrise, ψ in Eq. (11.5) is set to $90°$. Biologically significant times, especially for flowering and insect activity, begin in twilight hours just before sunrise and extend to just after sunset. Beginning and ending times for "civil twilight" are sometimes used to define these times of activity. Civil twilight is defined as beginning and ending when the sun is $6°$ below the horizon, so $\psi = 96°$.

Example 11.1. Find the sun zenith angle for Pullman, WA at 10:45 PDT on June 30. Also find the time of first twilight and the daylength.

Solution. Convert the time of observation to standard time by subtracting one hour and convert minutes to decimal hours, so $t = 9.75$ hrs. The calendar day for June 30 is $J = 181$. Pullman latitude is $46.77°$, and longitude is $117.2°$. The standard meridian is $120°$. The local meridian is therefore $2.8°$ east of the standard meridian, so $LC = 2.8° \div 15°/hr = 0.19$ hrs. From Eq. (11.4) or Table 11.1, $ET = -0.06$ hrs. Equation (11.3) then gives $t_o = 12 - 0.19 - (-0.06) = 11.87$ hrs. Declination, from Table 11.1 or Eq. (11.2) is $23.18°$. Substituting these values into Eq. (11.1) gives

$$\psi = \cos^{-1}\{\sin(46.77)\sin(23.18)$$
$$+ \cos(46.77)\cos(23.18)\cos[15(9.75 - 11.87)]\} = 34.9°.$$

The result is $34.9°$. The half daylength (including twilight), from Eq (11.6), is

$$\cos^{-1}\left[\frac{\cos(96) - \sin(23.18)\sin(46.77)}{\cos(23.18)\cos(46.77)}\right] = 128°.$$

Converting to hours gives $128° \times \frac{1\,hr}{15°} = 8.56$ hrs. The time of first twilight is $11.87 - 8.56 = 3.31$ hrs (solar time). The daylength is $2 \times 8.56 = 17.1$ hrs. The time of sunrise in PDT is 3.31 hrs $+ LC + ET + 1$ hr $= 4.44$ hrs (PDT).

11.2 Estimating Direct and Diffuse Short-wave Irradiance

Computation of the solar or shortwave component of the radiant energy budget of an organism requires estimates of flux densities for at least three radiation streams: direct irradiance on a surface perpendicular to the beam (S_p), diffuse sky irradiance on a horizontal plane (S_d), and reflected radiation from the ground (S_r). In addition to these, sometimes the beam irradiance on a horizontal surface S_b and the total (beam plus diffuse) irradiance of a horizontal surface S_t need to be known. S_t is sometimes reffered to as the global irradiance. These last two quantities are related to the first by

$$S_b = S_p \cos \psi \qquad (11.8)$$

and

$$S_t = S_b + S_d \qquad (11.9)$$

where ψ is the solar zenith angle.

Reflected radiation is the product of the average surface reflectance for the solar waveband and the total shortwave irradiance of the surface:

$$S_r = \rho_s S_t. \qquad (11.10)$$

The shortwave surface reflectance is called albedo. Typical albedos for several surfaces are given in Table 11.2. These values are influenced by amount of cover, color of soil or vegetation, and sun elevation angle, so the values in the table should be regarded as approximate. Tall canopies and water surfaces have reflectances that depend strongly on solar zenith angle. The values in Table 11.2 are for small (midday) zenith angles.

Though a number of models are available for estimating clear sky S_p and S_d with considerable accuracy (McCullough and Porter, 1971), they require data that are not generally available to the ecologist without special measurements, and are quite complicated to use. We use a simpler model based on Liu and Jordan (1960). We expect S_p to be a function of the distance the solar beam travels through the atmosphere, the transmittance of the atmosphere, and the incident flux density. A simple expression combining these factors is:

$$S_p = S_{po} \tau^m \qquad (11.11)$$

where S_{po} is the extraterrestrial flux density in the waveband of interest, normal to the solar beam. The term τ is the atmospheric transmittance and m is the optical air mass number, or the ratio of slant path length

TABLE 11.2. Shortwave reflectivity (albedo) of soils and vegetation canopies.

Surface	Reflectivity	Surface	Reflectivity
Grass	0.24–0.26	Snow, fresh	0.75–0.95
Wheat	0.16–0.26	Snow, old	0.40–0.70
Maize	0.18–0.22	Soil, wet dark	0.08
Beets	0.18	Soil, dry dark	0.13
Potato	0.19	Soil, wet light	0.10
Deciduous forest	0.10–0.20	Soil, dry light	0.18
Coniferous forest	0.05–0.15	Sand, dry white	0.35
Tundra	0.15-0.20	Road, blacktop	0.14
Steppe	0.20	Urban area (average)	0.15

through the atmosphere to zenith path length. For zenith angles less than 80°, refraction effects in the atmosphere are negligible, and m is given by:

$$m = \frac{p_a}{101.3 \cos \psi}. \qquad (11.12)$$

The ratio $p_a/101.3$ is atmospheric pressure at the observation site divided by sea level atmospheric pressure, and corrects for altitude effects. Equation (3.7) can be used to calculate this ratio. It can be shown that Eq. (11.11) is mathematically equivalent to Beer's law (Eq. (10.4)).

Liu and Jordan (1960) measured τ on clear days, and found values ranging from 0.75 to around 0.45 at two sites. When τ is lower than about 0.4, one would consider the sky to be overcast. Gates (1980) suggests values of τ between 0.6 and 0.7 to be typical of clear sky conditions. Values on the clearest days would be around 0.75.

Of the radiation that starts through the atmosphere, part reaches the ground as beam radiation (Eq. (11.11)), part is absorbed by the atmosphere, part is scattered back to space, and part is scattered downward toward the ground. The down scattered part is called the sky diffuse radiation. The actual amount of diffuse radiation reaching the ground is difficult to compute because it depends, in part, on the albedo of the ground. All else being equal, the sky is brighter when the ground is snow covered than it is when the ground is covered with dense, dark vegetation. Without getting into these complications, approximate values can be computed for sky diffuse radiation on clear days using an empirical equation adapted from Liu and Jordan (1960):

$$S_d = 0.3(1 - \tau^m)S_{po} \cos \psi. \qquad (11.13)$$

The airmass factor partially compensates for the effect of the cosine factor in Eq. (11.13), so that the diffuse radiation remains relatively constant throughout clear days. In fact, Peterson and Dirmhirn (1981) found that the ratio S_d/S_p is nearly constant on clear days. Figure 11.2 shows the beam, diffuse, and total radiation computed using Eqs. (11.11) and (11.13) for a clear atmosphere. Figure 11.3 shows these same radiation streams, but for a turbid atmosphere. Note that as the dust and haze increase, beam radiation is decreased and diffuse radiation increases.

11.3 Solar Radiation under Clouds

When clouds obscure the sun, $S_d = S_t$, since there is no beam radiation component. Empirical transmission coefficients have been worked out for various cloud types and used to determine the shortwave irradiance under clouds. Total shortwave irradiance is shown in Fig. 11.4 as a function of solar elevation angle for various cloud types. Clearly the presence of some kinds of clouds can cause widely fluctuating irradiance so the curves in Figure 11.4 are averages.

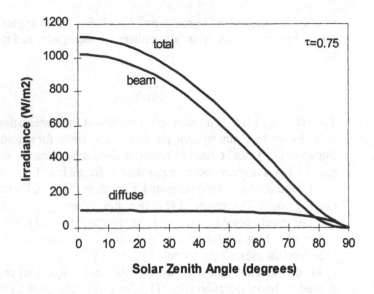

FIGURE 11.2. Beam (S_b), diffuse (S_d), and total solar radiation (S_t) as a function of zenith angle for a very clear sky.

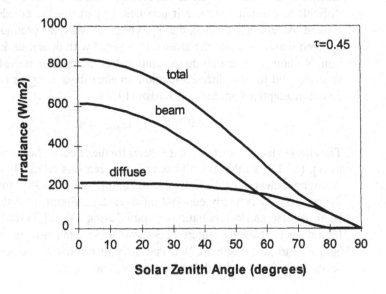

FIGURE 11.3. Similar to Fig. 11.2, but for turbid or polluted air.

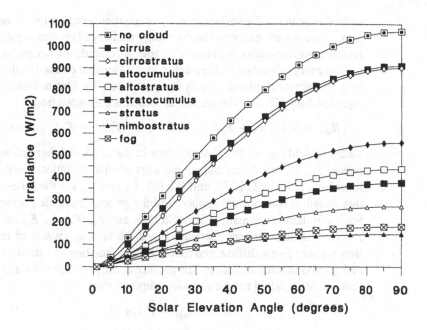

FIGURE 11.4. Solar irradiance under cloud cover for various cloud types and solar elevation angles (data from List, 1971).

Example 11.2. Find all components of the solar radiation over a grass surface on a clear day when the solar zenith angle is 30° and the altitude is 800 m.

Solution. The cosine of the zenith angle is needed in several places; it is cos(30) = 0.866. The atmospheric pressure is (Eq. (3.7)).

$$p_a = 101 \exp\left(\frac{-800}{8200}\right) = 91.6\,\text{kPa}.$$

The airmass number is (Eq. 11.12)

$$m = \frac{91.6}{101 \times 0.866} = 1.05.$$

The transmittance, τ^m is needed to compute both S_p and S_d. Assuming a value of 0.7 for τ gives $0.7^{1.05} = 0.69$. Now, using Eq. (11.11), $S_p = 1360\,\text{W m}^{-2} \times 0.69 = 938\,\text{W m}^{-2}$. Using Eq. (11.13), $S_d = 0.3 \times 1360\,\text{W m}^{-2} \times 0.866 \times (1 - 0.69) = 110\,\text{W m}^{-2}$. Using Eqs. 11.8 and 11.9, $S_t = 938\,\text{W m}^{-2} \times 0.866 + 110 = 922\,\text{W m}^{-2}$. Finally, the reflected radiation is obtained from Eq. (11.10) using the average grass albedo from Table 11.2: $S_r = 0.26 \times 922\,\text{W m}^{-2} = 240\,\text{W m}^{-2}$.

11.4 Radiation Balance

Part of the radiation incident on a leaf, an animal, a crop canopy, or a soil surface is absorbed at the surface. This energy can warm the surface or be dissipated to the environment by conduction, evaporation, or radiation. Since we are interested in the energy exchange between organisms and their surroundings, we need to be able to compute the net amount of radiant energy absorbed by the surface, and the amount emitted by the surface. The amount emitted is easily computed using the Stefan–Boltzmann equation from Ch. 10. The amount absorbed is computed from:

$$R_{abs} = \alpha_s(F_p S_p + F_d S_d + F_r S_r) + \alpha_L(F_a L_a + F_g L_g) \qquad (11.14)$$

where α_s and α_L are the absorptivities in the solar and thermal wavebands, S_p, S_d, and S_r are the components of solar radiation, computed from Eqs. (11.11), (11.13), and (11.10), L_a and L_g are the long-wave flux densities from the atmosphere and the ground (computed using the Stefan–Boltzmann equation from Ch. 10), and F_p, F_d, F_r, F_a, and F_g are view factors between the surface and the various sources of radiation; namely beam, diffuse, and reflected solar radiation and atmospheric and ground thermal radiation. The net flux density of radiant energy at a surface, often called net radiation is computed from:

$$R_n = R_{abs} - F_e \varepsilon_s \sigma T_s^4 \qquad (11.15)$$

where ε_s is the emissivity of the surface, T_s is the surface temperature (in kelvins), and F_e is the view factor between the entire surface of the object and the complete sphere of view. For convex surfaces, $F_e = 1$ except for the unusual case of an animal lying on the ground so that only a portion of its total surface area is emitting radiation. For complex surface shapes, F_e may be less than one because some of the surface "views" other parts of the surface of the same object or animal. Equation (11.15) expresses the radiation balance at the surface, and Eq. (11.14) gives the information needed to compute the radiation balance. In order to make the computation shown in Eq. (11.14), values are needed for the absorptivities and view factors.

11.5 Absorptivities for Thermal and Solar Radiation

According to Kirchhoff's law, given in Ch. 10, the absorptivity in a given waveband is equal to the emissivity in that waveband. The longwave absorptivity needed for Eq. (11.14) is therefore equal to the emissivity of the surface. In Ch. 10 we give a typical value for emissivities of natural surfaces of around 0.97. Table 11.3 gives measured values for leaves, animals, and various other surfaces. Note that, except for metal surfaces, the emissivities are around the 0.97 value used in Ch. 10. We therefore continue to use this value for emissivities of natural surfaces and for absorptivities of leaves and animals. Obviously a much lower value should be used for a metal surface. Note that a polished metal coating on a surface (gold, silver, or aluminum) can almost eliminate both the absorption and the emission of thermal radiation. This fact is used in the design of Thermos bottles. By silvering the glass surfaces of the bottle the emis-

TABLE 11.3. Long-wave or thermal emissivities (and absorptivities) for leaves, animals, and other surfaces

Surface	Emissivity	Surface	Emissivity
maize leaf	0.94	human skin	0.98
tobacco leaf	0.97	snowshoe hare	0.99
bean leaf	0.94	caribou	1.00
cotton leaf	0.96	gray wolf	0.99
sugar cane leaf	0.99	gray squirrel	0.99
poplar leaf	0.98	window glass	0.90–0.95
cactus	0.98	concrete	0.88–0.93
polished chrome	0.05	soil	0.93–0.96
bright aluminum foil	0.06	water	0.96

sivity is reduced from above 0.9 to below 0.05, thus almost eliminating radiative exchange between the inner and outer bottle surfaces.

In Ch. 10 we discuss the computation of absorptivities for radiation and indicate that the absorptivity for a particular source of radiation is the normalized integral of spectral absorptivity weighted by the spectral irradiance of the source. Figure 11.5 shows the spectral absorptivities of some leaf and animal surfaces in the shortwave region of the spectrum. While all of the surfaces show variation of absorptivity with wavelength, the leaf absorptivity changes dramatically between the visible and near infrared portions of the spectrum. In the visible, most of the radiation is absorbed, and is used to carry on photosynthesis. Absorption is somewhat lower in the green (around 0.55 μm) part of the spectrum, resulting in

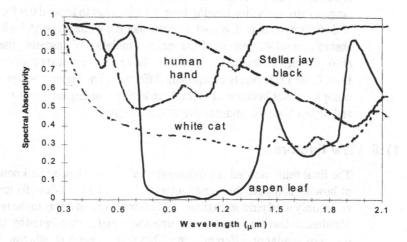

FIGURE 11.5. Spectral absorptivity of leaf, fur, feather, and skin surfaces over part of the solar spectrum (data from Gates, 1980, and Hall et al., 1992).

TABLE 11.4.   Shortwave absorptivities of leaves and animals (from Gates, 1980).

Leaves		Mammals	
silver maple	0.48–0.56	bison	0.78
american beach	0.47–0.52	wolf	0.80
sunflower	0.52–0.57	cat (white)	0.445
cottonwood	0.50	bobcat	0.70
cottonwood (yellow)	0.39	Reptiles	
Birds		alligator	0.90
Stellar's jay	0.88	lizard	0.90
sparrow (dorsal)	0.75	Humans	
quail (dorsal)	0.72	Eurasian	0.68–0.72
quail egg	0.18	Negroid	0.79
white swan	0.37		

the characteristic green color of vegetation, but overall, approximately 85 percent of the incident visible radiation is absorbed while about 15 percent of the NIR is absorbed. The NIR wavelengths are not useful for biochemical processes and are largely reflected or transmitted by the leaf.

Integration of data like that shown in Fig. 11.5, with weighting according to the solar spectrum (Fig. 10.5), results in the shortwave absorptivitiy for solar radiation. Table 11.2 shows solar reflectivities for various types of ground cover. Since all of the incident radiation is either reflected or absorbed, the absorptivity of these surfaces can be computed as $\alpha_s = 1 - \rho_s$ where ρ_s comes from Table 11.2. Representative animal and leaf absorptivities are given in Table 11.4. Gates (1980) gives a more comprehensive table.

It appears that shortwave absorptivities of leaves are around 0.5, so about half of the incident solar radiation is absorbed. Animals have a wide range of absorptivities ranging from 0.18 for eggs to around 0.9 for black or dark brown coats. Even white coats like the white swan and white cat absorb around 40 percent of the incident radiation. Comparing the solar absorptivities of leaves (~ 0.5) from Table 11.4 with canopies (~ 0.8) from Table 11.2 reveals a surprising difference. The higher absorptivity of canopies arises because of multiple reflections among leaves in a canopy and depends on the architecture of the canopy.

11.6 View Factors

The final thing needed for finding the radiation balance is a knowledge of how to compute the view factors in Eq. (11.14). "View factor" is a commonly used term in engineering heat transfer and refers to the fraction of radiation leaving one object of some shape that is intercepted by another object of similar or different shape. Thus if an object A is radiating and an object B is receiving some of that radiation, then the view factor would be expressed as F_{A-B}. This view factor is generally different from the view factor for radiation leaving object B and received by object A — F_{B-A}. In a simple conceptual model of outdoor radiation the sources would be the sun, sky, and ground. For our purposes, a diffuse source of radiation like the sky can be treated as a hemispherically-shaped object of exceedingly large radius. In engineering, view factors are computed between objects; whereas in environmental biophysics the interest is usually in view factors between objects and sources of radiation. By considering the example of a small sphere located inside of a large sphere, these two applications of view factors can be related to each other. The view factor from the small sphere to the large sphere is 1.0, because all the radiation leaving the small sphere is received by the large sphere. Alternatively, all the view of the small sphere is entirely occupied by the large sphere because the small sphere cannot view any of itself. In contrast, the view factor from the large sphere to the small sphere is given by the ratio of the sphere areas (A_{small}/A_{large}) so that all the radiation leaving the large sphere is not received by the small sphere but some is received by the large sphere itself. This fraction of radiation leaving the large sphere that impinges on other areas of the same large sphere is the view factor between the large sphere and itself ($1 - A_{small}/A_{large}$). As the small sphere, which is inside the large sphere, becomes smaller and smaller, the view factor from the large sphere to the small sphere becomes small; but the view factor between the small sphere and the large sphere remains 1.0. In environmental biophysics we are usually interested in the view factor between an object, such as the interior sphere in the example above which might be a bird flying, and the imaginary sphere surrounding that object representing the source of radiation, such as the sky and ground. Usually the sphere of view surrounding the object has several sources of radiation such as sky or ground, so the view of the object is divided up into the various components that represent the various sources of radiation; but the sum of all these view factors must always be 1.0. If the exterior (large) sphere in the example above is divided into two hemispheres, the view factor between the interior (small) sphere and the exterior (large) upper hemisphere is 0.5. Likewise the view factor between the interior (small) sphere and the lower exterior (large) hemisphere is 0.5. If the interior (small) sphere is replaced by an interior (small) cylinder or thin flat plate, the view factor between this cylinder or thin flat plate and either exterior hemisphere also is 0.5.

Frequently the thin flat plate case leads to the greatest confusion because the area of thin flat plates is almost always specified by the "one-sided" area or "silhouette" area by convention. You might normally consider a 3 in. × 5 in. card to have an area of 15 in.^{2} because you tend to use the convention of one-sided area. In this book, however, we consider such a card to have an area of 30 in.^{2} because we always use the area of the total surface (top and bottom). Because we generally are interested in the view factor between an object and radiation originating from some portion of a sphere, we omit the initial subscript denoting the object so that a view factor between a leaf and the sky is given by F_{sky} not $F_{leaf-sky}$.

Although the sky usually is considered to be a hemisphere, it need not be so; for example, the view factor between a leaf and the sky may be less than one for the bottom of a mountain gorge.

In engineering, view factors usually are only used for diffuse radiation; however, in environmental biophysics we want to use them for beam and diffuse radiation so we expand the definition somewhat. Here we define the view factor as the average flux density over the entire surface of some object of interest divided by the flux density on a flat absorbing surface facing the source. The surfaces of interest to us are soil surfaces, plant canopies, individual leaves, and animals. When we say average flux density here, we mean the number of Watts of energy absorbed, averaged over the entire surface area of the object in question. For beam radiation this average ratio is easily obtained (at least in principle). It is equal to the ratio of the projected area in the direction of the solar beam (A_p) to the total surface area of the object (A). You might picture holding a piece of paper perpendicular to the beam of solar radiation and tracing the shadow of the object (animal, leaf, etc.) on the paper. The area of the shadow is the A_p. Thus for a flat horizontal leaf with the sun at zenith angle ψ, $F_p = 0.5 \cos \psi$ because only one side of the leaf faces the sun. View factors for the sky and ground are for diffuse radiation and was discussed earlier. If the source of radiation (sky or ground) is assumed to be isotropic (same intensity in all directions), then determining the view factor essentially means that each point on the sky or ground is considered as a source and each point on the object as a receiver. Each portion of the source is weighted by the solid angle it subtends with each portion of the receiver, each portion of the receiver is weighted by the fraction of the entire receiver area it occupies, and the incident radiation is multiplied by the cosine of the angle between the received-radiation direction and a normal to the surface at the point of absorption. Integrating over the entire solid angle of the source and area of the receiver gives the view factor. For a flat horizontal plate, a sphere, or a cylinder under a diffuse sky and over a diffusely reflecting soil surface, the view factor between the object (plate, sphere, or cylinder) and the sky hemisphere is 0.5 and the view factor between the object and the ground is 0.5. Of course this means that the view factor between the object and its entire view is 1.0. If the object were at the bottom of a deep canyon, then the view factor between the object and the sky might be 0.4 and between the object and the ground be 0.6 because some of the view of the top of the object is occupied by canyon walls.

Now, we specifically consider view factors for soils and plant canopies, animals and individual leaves. For a soil surface or a plant canopy $F_p = \cos \theta$, $F_d = F_a = F_e = 1$, and $F_r = F_g = 0$, where θ is the angle between the solar beam and a normal to the plane of the soil or canopy. For a horizontal surface $\theta = \psi$, the zenith angle. When the surface has slope, θ can be computed from

$$\cos \theta = \cos \gamma \cos \psi + \sin \gamma \sin \psi \cos(AZ - AS) \qquad (11.16)$$

where γ is the inclination angle of the surface, ψ is the zenith angle of the sun, AZ is the azimuth angle of the sun with respect to due south, and AS is the aspect angle (angle between south and the projection onto the horizontal of the normal to the inclined surface) of the surface. If the surface is sloped, F_d and F_a usually are less than 1.0; typically $F_d = F_a = (1 + \cos\gamma)/2$.

It may seem strange that F_r and F_g are zero for a plant canopy since the leaves do receive reflected and emitted radiation from the soil and from lower layers in the canopy. The important thing to remember here is that we are not trying to deal with the details of leaf processes when we compute the absorption of radiation by a canopy. We imagine that we are far enough from the canopy so that we can treat it as a single, flat surface which absorbs and emits radiation. Thus we consider the canopy to be an object with only one side; a practical impossibility, but useful and consistent conceptually. Later we deal with the details of radiative exchange by canopy elements.

As stated earlier, F_p for an animal is just equal to A_p/A, the ratio of projected area perpendicular to the solar beam to total animal area. Figure 11.6 shows this ratio for several objects which approximate the shapes of animals. To use Fig. 11.6, one simply determines the angle between the longitudinal axis of the animal and the solar beam, the general shape of the animal, and the ratio of length to diameter. It appears that the view factor should fall in the range 0.1 to 0.3. A sphere has a view factor for beam radiation of 0.25.

If the effect of shadows are ignored, the diffuse view factors for an animal suspended above the ground hemisphere and below a sky hemisphere are $F_d = F_r = F_a = F_g = 0.5$ and $F_e = 1$. Both the sky and the ground "see" more than half the body surface area of the animal, but the cosine weighting results in the view factor being 0.5. If the animal is lying on the ground, a large fraction of its surface is not accessible to radiation, and F_e, F_r and F_g must be adjusted accordingly.

The view factors for a single leaf are similar to those for an animal. The view factors for a leaf suspended over the ground and under the sky hemisphere are $F_p = 0.5\cos\theta$, $F_d = F_r = F_a = F_g = 0.5$, and $F_e = 1$. Equation (11.16) is used to compute θ.

Example 11.3. Find the net radiation for the grass surface in Example 11.2 if the air temperature is 30° C and the grass temperature is 35° C.

Solution. From Example 11.2, $S_p = 938$ W m^{-2}, $S_d = 110$ W m^{-2}, and $S_r = 240$ W m^{-2}. From Table A.3, the black body emittances for 30° C and 35° C are 479 and 511 W m^{-2}, and the clear sky emissivity at $T_a = 30$° C is 0.85. The sky thermal radiant emittance is therefore $L_a = 0.85 \times 479 = 407$ W m^{-2}. The ground thermal radiant emittance is $\varepsilon_s \sigma T_s^4 = 0.97 \times 511 = 496$ W m^{-2}. The shortwave absorbtivity is equal to $1 - \rho_s$. For grass $\rho_s = 0.26$ (Example 11.2) so $\alpha_s = 0.74$.

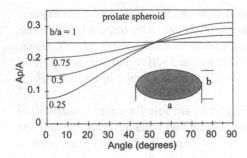

$$\frac{A_p}{A} = \frac{\sqrt{1+(x^2-1)\cos^2\theta}}{2x + \dfrac{2\sin^{-1}\sqrt{1-x^2}}{\sqrt{1-x^2}}} \quad ; \quad x = b/a$$

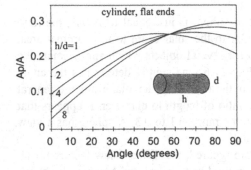

$$\frac{A_p}{A} = \frac{\cos\theta + \dfrac{4h\sin\theta}{\pi d}}{2 + \dfrac{4h}{d}}$$

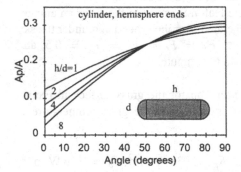

$$\frac{A_p}{A} = \frac{1 + \dfrac{4h\sin\theta}{\pi d}}{4 + \dfrac{4h}{d}}$$

FIGURE 11.6. Ratios of shadow area on a surface perpendicular to the solar beam to total surface area for three simulated animal shapes. The angle indicated is the angle between the solar beam and the longitudinal axis of the solid.

Now, using Eq. (11.14), with $F_p = \cos \psi = \cos 30 = 0.866$ (the sun zenith angle was given as 30°); $F_d = F_a = 1$, and $F_r = F_g = 0$, gives $R_{abs} = 0.74 \times (0.866 \times 938 \text{ W m}^{-2} + 1 \times 110 \text{ W m}^{-2}) + 0.97 \times (1 \times 407 \text{ W m}^{-2}) = 1077 \text{ W m}^{-2}$. The net radiation is (Eq. (11.15)): $R_n = 1077 \text{ W m}^{-2} - 496 \text{ W m}^{-2} = 581 \text{ W m}^{-2}$.

Example 11.4. If a sparrow were standing on the grass, what would its R_{abs} be?

Solution. From Table 11.4, the sparrow absorptivity is 0.75. Consulting Fig. 11.6 it can be seen that the value of F_p could range from about 0.15 to 0.35 depending on the orientation of the bird to the solar beam. We calculate the absorbed radiation for both extreme values, using the values from Example 11.3.
Smallest R_{abs}:

$$R_{abs} = 0.75 \times (0.15 \times 938 + 0.5 \times 110 + 0.5 \times 240)$$
$$+ 0.97 \times (0.5 \times 407 + 0.5 \times 496)$$
$$= 675 \text{ W m}^{-2}.$$

Largest R_{abs}:

$$R_{abs} = 0.75 \times (0.35 \times 938 + 0.5 \times 110 + 0.5 \times 240)$$
$$+ 0.97 \times (0.5 \times 407 + 0.5 \times 496)$$
$$= 815 \text{ W m}^{-2}.$$

Clearly, the bird has access to a wide range of absorbed radiation environments, just by choosing its orientation with respect to the sun.

Example 11.5. If a single flat leaf were suspended horizontally over the same grass surface, what would its absorbed radiation be?

Solution. From Table 11.4, the absorptivity for the leaf is around 0.5. Since the leaf is horizontal, $F_p = 0.5 \times \cos \psi = 0.433$. All of the other view factors are 0.5, and $F_e = 1$. The absorbed radiation is therefore:

$$R_{abs} = 0.5 \times (0.433 \times 938 + 0.5 \times 110 + 0.5 \times 240)$$
$$+ 0.97 \times (0.5 \times 407 + 0.5 \times 496)$$
$$= 729 \text{ W m}^{-2}.$$

References

Coulson, K. L. (1975) Solar and Terrestrial Radiation. Academic Press, New York.

Gates, D. M. (1962) Energy Exchange in the Biosphere. New York: Harper and Row.

Gates, D. M. (1965) Radiant energy, its receipt and disposal. Meteor. Monogr, 6:1–26.

Gates, D. M. (1980) Biophysical Ecology. New York: Springer Verlag.

Hall, F.G., K.F. Huemmerich, D.E. Strebel, S.J. Goetz, J.E. Nickeson, and K.D. Woods, (1992) Biophysical, Morphological, Canopy Optical Property, and Productivity Data From Superior National Forest. NASA Tech. Mem. 104568, Goddard Space Flight Center, Greenbelt, MD 20771.

List, R. J. (1971) Smithsonian Meterological Tables, 6th ed. Washington D. C.: Smithsonian Institution Press.

Liu, B. Y., and R. C. Jordan (1960) The interrelationship and characteristic distribution of direct, diffuse, and total solar radiation. Solar Energy 4:1–19.

McCullough, E. C., and W. P. Porter (1971) Computing clear day solar radiation spectra for the terrestrial ecological environment. Ecology 52:1008–1015

Monteith, J. L. and M. H. Unsworth (1990) Principles of environmental Physics, second edition. London: Edward Arnold.

Munn, R. E. (1966) Descriptive Micrometerology. New York: Academic Press.

Peterson, W. A., and I. Dirmhirn. (1981.) The ratio of diffuse to direct solar irradiance (perpendicular to the sun's rays) with clear skies - a conserved quantity throughout the day. J. Appl. Meteorol. 20:826–828.

Weiss, A. and J. M. Norman. (1985.) Partitioning solar radiation into direct and diffuse, visible and near-infrared components. Agric. Forest. Meteorol. 34:205–213.

Problems

11.1. Compare R_{abs} for an animal under a clear night sky and a completely overcast cloudy night sky. Assume $T_a = T_g = 20°$ C.

11.2. Compute R_{abs} for your hand suspended horizontally over a dry soil surface on a clear day. Assume $T_a = 25°$C, $T_g = 45°$C, $\psi = 40°$, and $\alpha_s = 0.65$. Estimate values not given and state assumption for making estimates. Estimate the total watts absorbed by your hand.

11.3. Give all components of the radiant energy budget, and compute the net radiation for a bare soil surface under a clear sky on April 10 at 1100 hrs. solar time. Latitude is 37° and elevation is 1200 m. Air temperature is 18° C and soil surface temperature is 22° C.

11.4. Make the same computations as in problem 11.3, but for an overcast sky.

11.5. Find the elevation, zenith, and azimuth angles of the sun at 600 hrs (solar time) on June 21, at a latitude of 48° and longitude of 95°.

11.6. Find the daylength and time of sunrise and sunset using the data in problem 11.5.

11.7. Estimate the duration of civil twilight in hours for Example 11.1.

Animals and their Environment 12

The principles discussed thus far become more meaningful as they are applied to problems in nature. The first problem considered is that of describing the fitness of the physical environment for survival of an animal whose requirements we specify. Survival of the animal can depend on many factors; we consider only those related to maintaining body temperature within acceptable limits and those related to maintaining proper body water status. Even these aspects are only discussed to a limited extent. For example, maintenance of body temperature in endotherms (animals which maintain body temperature through internal metabolic heat production) involves production of metabolic heat. Stored chemical energy from the animal's food is used to produce the heat, so availability of food in the environment could be construed as part of the animal's physical environment. Food availability does not enter into our discussions in this way, but we do compute the amount of food an endotherm needs in order to maintain constant body temperature. Food requirements are of interest to those modeling ecosystems as well as those managing range lands for wild or domestic animals.

12.1 The Energy Budget Concept

The question of whether or not an animal can maintain its body temperature within acceptable limits can be stated in another way which makes it more amenable to analysis. It can be asked whether heat loss can be balanced by heat input and production at the required body temperature. We are well prepared to describe heat inputs and heat losses for a system, so the problem is easily solved, at least in principle. An equation stating that heat inputs minus heat losses equals heat storage for a system is called an energy budget equation. As an example of an energy budget, consider a representative unit area of the surface of an animal that is exposed to the atmosphere. The energy budget of this surface is the sum of the heat inputs and losses. Thus:

$$R_{abs} - L_{oe} + M - \lambda E - H - G - q = 0 \qquad (12.1)$$

where R_{abs} is the flux density of absorbed radiation, L_{oe} is the flux density of outgoing, emitted radiation from the surface, M is the rate of metabolic heat production per unit surface area, λE is the latent heat loss from evaporation of water, H is the rate of sensible heat loss, G is the rate of heat loss to the substrate by conduction, and q is the rate of heat storage in the animal per unit surface area.

Initially, we concern ourselves only with steady-state conditions for which the heat storage rate q is zero. The rate of heat storage is equal to the heat capacity of the animal multiplied by the rate of change of body temperature, so if the heat storage rate is zero, the rate of change of body temperature must also be zero. For simplicity, we also assume $G = 0$.

The emitted radiation $[L_{oe} = \varepsilon_s \sigma T_s^4]$ and sensible heat $[H = c_p g_H (T_s - T_a)]$ terms both involve the surface temperature of the animal. It is always possible to set a value for surface temperature which balances Eq. (12.1), but that temperature may be too high or too low for the animal to remain alive. If body temperature and metabolic rate are specified, then Eq. (12.1) can be used to find environments that are energetically acceptable (R_{abs} and T_a that will balance the energy budget). On the other hand, we could measure or estimate R_{abs}, T_a, and T_b and compute M. Knowing M, we can specify food needs for thermoregulation in a given climate.

Equation (12.1) is not very useful as it stands because of its strong dependence on surface temperature, a quantity that is hard to estimate, or even to measure. Body temperature, at least for endotherms, is easily estimated since it is under tight metabolic control. This fact can be used to eliminate surface temperature from the energy balance equation. Figure 12.1 shows the assumptions we make about the source (M) and sink (λE) of heat, and the resistances to heat flow from the body core to the environment. In a nonsweating animal, much of the latent heat loss is through breathing, or panting, and the remainder is from beneath the coat, which generally has a much higher resistance than the tissues. It is therefore justified to lump latent heat loss with metabolic heat production and place them at the body core. Later we do a more complete analysis which does not restrict the location of the latent heat loss. It is often useful to combine coat and tissue conductance into a whole body conductance:

$$g_{Hb} = \frac{g_{Ht} g_{Hc}}{g_{Ht} + g_{Hc}}. \tag{12.2}$$

Since all of the heat from the body core flows through g_{Hb}, we can write

$$M - \lambda E = c_p g_{Hb}(T_b - T_s) \tag{12.3}$$

where c_p is the specific heat of air. Equation (12.1) can now be rewritten explicitly showing the surface temperature dependence:

$$R_{abs} - \varepsilon_s \sigma T_s^4 + M - \lambda E - c_p g_{Ha}(T_s - T_a) = 0. \tag{12.4}$$

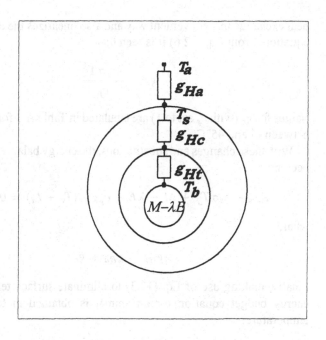

FIGURE 12.1. Diagram of heat production and loss in an non-sweating animal.

By combining Eq. (12.3) and Eq. (12.4), we can eliminate the surface temperature and have an energy balance equation in terms of body temperature and whole body conductance.

Before proceeding with the derivation of the energy balance equation, we briefly consider an algebraic manipulation which linearizes the surface emittance term, where surface temperature is raised to the fourth power. T_s can be written as $T_a + \Delta T$, where $\Delta T = T_s - T_a$. Now the binomial expansion is used to obtain:

$$(T_a + \Delta T)^4 = T_a^4 + 4T_a^3\Delta T + 6T_a^2\Delta T^2 + 4T_a\Delta T^3 + \Delta T^4. \quad (12.5)$$

A calculator can be used to verify that the terms in Eq. (12.5) with powers of ΔT greater than one are negligibly small for values of ΔT up to tens of degrees. Therefore T_s^4 can be approximated as $T_a^4 + 4T_a^3\Delta T$. The approximation is almost exact if, instead of using the cube of the air temperature, the cubed average of surface and air temperature is used.

Using this approximation, the surface emittance term in Eq. (12.4) can be written as:

$$\varepsilon_s\sigma T_s^4 = \varepsilon_s\sigma T_a^4 + 4\varepsilon_s\sigma T_a^3(T_s - T_a) = \varepsilon_s\sigma T_a^4 + c_p g_r(T_s - T_a). \quad (12.6)$$

Here we have defined a radiative conductance. For an animal in an enclosure the net exchange of thermal radiation between the walls of the enclosure and the animal is directly proportional to the difference between wall temperature and animal surface temperature and also directly proportional to the radiative conductance. This conductance therefore allows the combination of thermal radiative exchange with convective heat exchange in a convenient way and also linearizes the energy balance equation. From Eq. (12.6) it is seen that

$$g_r = \frac{4\varepsilon_s \sigma \mathbf{T}^3}{c_p}. \tag{12.7}$$

Values for g_r (with ε_s set to 1) are tabulated in Table A.3 for temperatures between -5 and 45°C.

With these changes and substitutions, the energy balance equation now becomes:

$$R_{abs} - \varepsilon_s \sigma T_a^4 + M - \lambda E - c_p g_{Hr}(T_s - T_a) = 0 \tag{12.8}$$

where

$$g_{Hr} = g_{Ha} + g_r. \tag{12.9}$$

Finally, making use of Eq. (12.3) to eliminate surface temperature the energy budget equation for an animal is obtained in terms of body temperature:

$$R_{abs} - \varepsilon_s \sigma \mathbf{T}_a^4 \\ + (M - \lambda E)(1 + g_{Hr}/g_{Hb}) - c_p g_{Hr}(T_b - T_a) = 0. \tag{12.10}$$

One final simplification allows the energy balance equation to be written in a particularly useful form. Animal metabolism is often studied inside chambers where air and wall temperatures are equal, where the flux density of shortwave radiation is negligible, and where wall emissivities are high. Such a chamber could be called a blackbody enclosure. If the radiation balance equation (Eq. (11.14)) is looked at for an animal in such an enclosure, it is seen that $R_{abs} = \varepsilon_s \sigma \mathbf{T}_a^4$. We call the temperature of such a blackbody enclosure the operative temperature, with the symbol T_e. Later we relate the operative temperature to outdoor radiation and temperature conditions, but for now the use of operative temperature allows us to eliminate radiation terms from the energy balance equation. The operative temperature form of Eq. (12.10), with some rearrangement of terms is:

$$M - \lambda E = \frac{c_p g_{Hr}(T_b - T_e)}{1 + g_{Hr}/g_{Hb}} = \frac{c_p(T_b - T_e)}{r_{Hb} + r_{Hr}}. \tag{12.11}$$

The second equation, in resistance form, is the more familiar form, but we continue to use conductances here. For now T_e can be thought of as the air temperature of a normal room in which the air and wall temperatures are equal. Equation (12.11) simply shows the relationship between temperature, resistance, metabolic rate, and latent heat loss for an animal. It is extremely useful for analyzing animal–environment interaction, but before going farther we need to consider some aspects of animal biology to get values for M, λE, and g_{Hb}.

12.2 Metabolism

The energy budget equation requires metabolic rate in terms of average energy supplied to the animal's surface per unit area. Physiologists generally measure metabolic rates per unit mass of animal. Control of body temperature for homeotherms and maintenance of body functions in all animals requires a minimal or basal metabolic rate. This basal rate (Watts) can be approximated for a wide variety of animals by the equation:

$$B_m = Cm^{3/4} \qquad (12.12)$$

where m is the animal's mass (kg) and C is a constant between three and five for endotherms and around five percent of this value for poikilotherms at 20°C. Figure 12.2 shows this relationship for a wide range of animal sizes.

An approximate relation between surface area (m^2) and mass (kg) is:

$$A = 0.1m^{2/3}. \qquad (12.13)$$

The uncertainty in the exponents of Eqs. (12.12) and (12.13) are large enough that, for many practical purposes, they may be taken as being the same. Thus the basal metabolic rate per unit animal surface area, $M(= B/A)$, is relatively independent of animal size. Typical values for M_b (the basal metabolic rate per unit area) in endotherms range from 30 to 50 W/m^2.

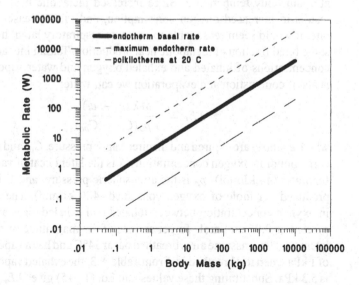

FIGURE 12.2. Relation between body mass and basal metabolic rate of poikilotherms at 20 C, basal metabolic rate of endotherms at 39 C, and maximum aerobic metabolic rate.

Metabolic rate increases with animal activity. This can be accounted for in the energy budget in one of two ways. If about 30 percent efficiency is assumed for conversion of chemical energy to work in animals, then for each unit of work done there will be about two units of heat produced. If it is known how much work is done, the metabolic production of heat can be calculated. The other method is somewhat simpler. As a rule of thumb, the maximum aerobic metabolic rate an animal can sustain can be assumed to be about ten times the basal rate (Fig. 12.2). If the animal's activity can be estimated as a percent of maximum (say from oxygen consumption measurements or running speed compared to maximum) the metabolic contribution can be estimated from:

$$M = M_b \left(1 + \frac{9\alpha}{\alpha_M} \right) \tag{12.14}$$

where α is the animal's activity and α_M is the maximum sustainable activity. If M_b is 50 W m^{-2}, M will vary between 50 and 500 W m^{-2}.

12.3 Latent Heat Exchange

Evaporation of water from the respiratory tract and from the skin result in latent heat loss from the animal. The total latent heat loss, needed for the energy budget equations, is the sum of the respiratory and skin latent heat losses. Respiratory loss is a direct result of the air exchange for breathing. Skin water loss was already treated in detail in Chs. 6 and 7.

In respiratory evaporation, air is breathed in at ambient vapor pressure and breathed out at the saturation vapor corresponding to the temperature of the nasal passages. In most species the nasal passages are maintained at about body temperature. Since increased metabolic heat production results in increased oxygen consumption, and this increases breathing rate, it would seem reasonable to compute respiratory latent heat loss as some fixed fraction of metabolic heat production. Taking into account the concentrations of inhaled and exhaled oxygen and water vapor, and the heats of combustion and evaporation we can write:

$$\lambda E_r = \frac{M \lambda (e_e - e_i)}{\Gamma p_a (C_{oi} - C_{oe})} \tag{12.15}$$

where e_e and e_i are expired and inspired vapor pressure, C_{oe} and C_{oi} are the corresponding oxygen concentrations, λ is the latent heat of vaporization for water (44 kJ/mol), p_a is the atmospheric pressure, and Γ is the heat produced per mole of oxygen consumed (480 kJ/mol). The difference in oxygen concentration between inhaled and exhaled air is around five percent or 0.05 mol/mol. To get an idea of the magnitude of respiratory latent heat loss, assume air is breathed out at 34°C and has a vapor pressure of 1 kPa when it is breathed in. From Table A.3, the exhaled vapor pressure is 5.3 kPa. Substituting these values into Eq. (12.15) gives $\lambda E_r = 0.1$ M.

Some animals with small nasal passages exhale air at temperatures well below body temperature. Figure 12.3 compares exhaled air temperatures for several bird species with values for humans and for kangaroo rats. The

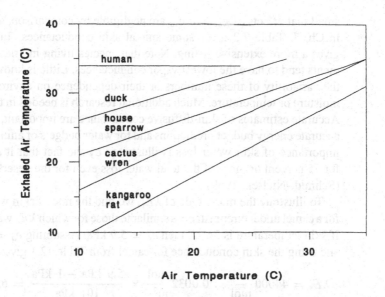

FIGURE 12.3. Temperature of exhaled air as a function of air temperature for several species (after Schmidt-Nielsen, 1972).

exhaled air temperature for the kangaroo rat is lower than air temperature, and approaches wet bulb temperature.

The appropriate value of e_e for animals which exhale air at temperatures lower than body temperature is the saturation vapor pressure at exhaled air temperature from Fig. 12.3. To see how effective this is for water conservation, compute λE_r for the kangaroo rat at 20°C. From Fig. 12.3, $T_s = 18°$ C, so $e_e = 2.1$ kPa. If the other values are as in the previous example, then $\lambda E_r = 0.02$ M, rather than 0.1 M. This is only 25 percent of the respiratory water loss per unit area of a human under similar conditions. The resulting water conservation is important for survival of kangaroo rats in their arid habitat.

Cutaneous latent heat loss is discussed in Chs. 6 and 7. The general equation is

$$\lambda E_s = \frac{g_v \lambda (e_s - e_a)}{p_a} \qquad (12.16)$$

where

$$g_v = \frac{1}{\frac{1}{g_{vs}} + \frac{1}{g_{vc}} + \frac{1}{g_{va}}};$$

g_{vs}, g_{vc}, and g_{va} are the conductances to vapor diffusion through the skin, coat, and boundary layer, and e_s and e_a are the vapor pressures at the subcutaneous (saturated) evaporating surface, and in the atmosphere. For animals with moist skins (earthworms, snails, and amphibians) g_{vc} and g_{vs} are large so the controlling conductance for water loss is the boundary layer conductance. For nonsweating animals, g_{vs} is often so small that effects of g_{vc} and g_{va} are negligible by comparison, as shown in Ch. 7. Table 7.2 gives some animal skin conductances. Table 12.1 gives a more extensive listing. Note that species living in arid environments tend to have the lowest vapor conductances. Little is known about the variability of these numbers or their dependence on environmental moisture or temperature. Much additional research is needed in this area. Accurate estimates of skin-diffusive conductance are important, both for accurate energy budget predictions and for water budgets of animals. The importance of skin water loss is illustrated by the fact that it accounts for 75 percent or more of the total water loss even for the desert tortoise (Schmidt-Nielsen, 1969).

To illustrate the magnitude of λE_s, we find the rate of skin water loss for a camel under circumstances similar to those for which λE_r was found. If skin temperature is 36°C, then $e_s = 5.9$ kPa. Assuming $e_a = 1$ kPa, and using the skin conductance for camel from Table 12.1 gives

$$\lambda E_s = 44000 \, \frac{J}{mol} \times 0.0032 \, \frac{mol}{m^2 s} \times \frac{5.9 \text{ kPa} - 1 \text{ kPa}}{101 \text{ kPa}} = 6.9 \, \frac{W}{m^2}.$$

The effect of coat and boundary layer conductance have been ignored, but their effect is small when the skin conductance is so low. If we assume $\lambda E_r = 0.1M$ and $M = 50$ W/m², then $\lambda E_r = 5$ W/m². The skin latent heat loss is larger than this value and, in fact, makes up about 58 percent of the total. The total latent heat loss is around 20 percent of M. These percentages are probably fairly typical for resting endotherms that are not heat-stressed. For poikilotherms under similar conditions one typically assumes $M = \lambda E$.

As the animal becomes heat-stressed, latent heat loss increases, generally by some active process such as sweating or panting. There is no general approach to the calculation of latent heat loss under these conditions since animal responses are so varied. The approach would need to be fitted to the particular species being studied. In Ch. 13 we look at

TABLE 12.1.   Skin conductance to vapor for non-heat stressed animals

Mammals	mmol m^{-2} s^{-1}	Reptiles (cont.)	mmol m^{-2} s^{-1}
white rat	10.6	gopher snake	1.0
human	5.4	chuckawalla	0.34
camel	3.2	desert tortoise	0.34
white footed mouse	3.0	**Birds**	
spiney mouse	2.8	sparrow	5.4
Reptiles		budgerigar	4.9
caiman	7.5	zebra finch	4.1
water snake	4.7	village weaver	3.3
pond turtle	2.8	poor-will	3.1
box turtle	1.3	roadrunner	2.4
iguana	1.2	painted quail	2.1

latent heat loss by sweating, but do not otherwise treat water loss under heat stress.

Example 12.1. White crown sparrows migrate over long distances. In flight, both water and energy are expended. If an average sparrow weighs 27 g, and can store 4 g of fat and 4 g of water (including the water obtained from metabolism), will its range be limited by stored energy or stored water? Assume the metabolic rate for flight is $6M_b$ and the energy content of fat is 40 kJ/g. Also assume $T_a = 10°$ C, $T_s = 38°$ C, and $e_a = 1$ kPa.

Solution. The energy-limited flight time is equal to the total energy available divided by the rate of energy consumption. The rate of energy consumption is $6M_b$. Assuming $M_b = 50$ W/m^2, $M = 300$ W/m^2. From Eq. (12.13), the sparrow area is $A = 0.1 \times (0.027)^{2/3} = 0.009$ m^2. The energy limited flight time is

$$t_{energy} = \frac{4g \times 40000 \frac{J}{g}}{300 \frac{J}{m^2s} \times 0.009 m^2} = 59,259 \text{ s} = 16.4 \text{ hour.}$$

Water loss is from the skin and from the respiratory tract. The vapor pressure of the air is given, and the vapor pressure at the evaporating surfaces is $e_s(38) = 6.63$ kPa. Assuming a flight altitude of 1000 m, then $p_a = 89$ kPa. Again we ignore effects of coat and boundary layer on vapor conductance, and use the sparrow skin conductance from Table 12.1. Using these numbers, the rate of skin water loss is

$$E_s = 0.0054 \frac{mol}{m^2s} \times \frac{6.63 \text{ kPa} - 1 \text{ kPa}}{89 \text{ kPa}} = 0.00034 \frac{mol}{m^2s}.$$

Equation (12.15) can be used to get the respiratory water loss. The inspired air is at the vapor pressure of the atmosphere and the expired air is saturated at the expired air temperature. From Fig. 12.3, the sparrow expired air temperature at 10°C ambient is 21°C. The saturation vapor pressure is 2.49 kPa. The rate of respiratory water loss is therefore

$$E_r = \frac{300 \frac{J}{m^2s} \times \frac{2.49 \text{ kPa} - 1 \text{ kPa}}{89 \text{ kPa}}}{480000 \frac{J}{mol} \times 0.05} = .00021 \frac{mol}{m^2s}.$$

The total rate of water loss is the sum of respiratory and skin loss, or 0.00045 mol m^{-2} s^{-1}. The stored water is 4 g $\div$ 18 g/mol = 0.22 mol. The water limited flight time is

$$t_{water} = \frac{0.22 \text{ mol}}{0.00045 \frac{mol}{m^2s} \times 0.009 m^2} = 54321 \text{ s} = 15 \text{ hours.}$$

With this set of assumptions, it appears that water and energy are about equally limiting. The assumptions were not arbitrary, but were taken from literature on white crown sparrows, so the conclusions are probably not far off. It should not be a surprise that the two limitations would be that well matched, since any unnecessary weight would also limit the range of the bird.

12.4 Conduction of Heat in Animal Coats and Tissue

The conduction of heat from the animal core to the environment is first through the vascularized tissues under the skin, then through the coat, and finally through the boundary layer to the surrounding air. Heat transfer from the body core to the skin surface of an animal depends on blood flow and is subject to regulation, within limits, by vasoconstriction or vasodilation. The regulation is important in control of body temperature. Table 12.2 gives maximum and minimum values of average tissue conductance for several species. These conductances, and their range of variation, would appear incapable of having much effect on overall heat loss from animals with coats because they are so large in comparison with coat conductances. Their important effect, however, is probably in controlling the surface temperature of poorly insulated appendages, which also have small characteristic dimensions and therefore large boundary layer conductances.

The conductance of animal coats is normally much lower than the tissue conductance, and is therefore the limiting conductance controlling heat loss. Figure 12.4 shows conductance for pieces of fur under laboratory conditions. In Ch. 7 conductances for layers of still air are computed. Since heat transport in animal coats can be by conduction through the air, by longwave radiative transport, and possibly by free convection, the conductance of air sets the lower limit for coat conductance. The coats in Fig. 12.4 follow the air conductance line reasonably well and are well below the line for radiative conductance in free space. It is interesting that coat conductance appears to stay fairly constant at around 40 to 50 mmol m^{-2} s^{-1} for coats thicker than 3 cm, no matter how thick the coat is.

The radiative conductance of a coat depends on the average distance radiation can travel within the coat (Cena and Monteith, 1975). The shorter the radiation path length the lower the radiative conductance. This dif-

TABLE 12.2.   Thermal conductance of peripheral tissue of animals (from Monteith and Unsworth, 1990; and Kerslake, 1972)

Animal	Vasoconstriction conductance (mol m^{-2} s^{-1})	Vasodilation conductance (mol m^{-2} s^{-1})
steer	0.24	0.83
calf	0.38	0.83
pig (3 months)	0.42	0.69
down sheep	0.46	1.4
human	0.46	2.8

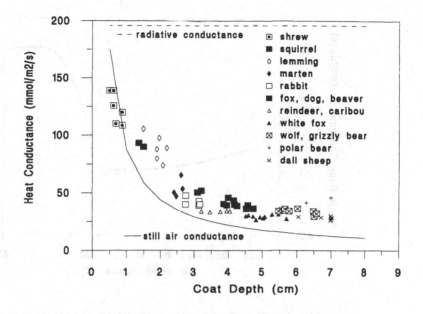

FIGURE 12.4. Coat conductance of animal fur compared to conductance of an equivalent thickness of still air and radiative conductance of open space.

ference in radiative path length is the main factor determining the quality of insulation in both homes and outdoor clothing. From Fig. 12.4 it appears that these coats are surprisingly effective at minimizing radiative transport within them.

As with the tissue conductance, getting an overall picture of coat effects on thermoregulation by looking only at conductances of pieces of fur is difficult. Coat depth varies from point to point, and an average thermal conductance for the entire body is needed. For this reason, conductances determined on live animals are likely to be more useful than those estimated on portions of animal coats. Calder and King (1974) give a relationship for the minimum conductance of birds, based on measurements of metabolic rate. It is

$$g_{Hb,\min} = 0.06 \ m^{-0.15} \text{mol m}^{-2}\text{s}^{-1} \qquad (12.17)$$

where m is the body mass in kg. In the absence of other information, minimum conductance could be computed from Eq. (12.17), or Fig. 12.4 could be used to estimate the minimum conductance for the animal assuming blood flow is restricted to the best insulated part of the body. The maximum conductance is achieved by shunting blood to poorly insulated appendages. Animals can often increase conductance by a factor of about three times the minimum by doing this. Figure 12.5 is a dramatic example of the range of conductance which can be achieved by an animal. Note that, for a resting white crown sparrow, minimum conductance occurs between about 10 and 25° C. By about 45°C the conductance has tripled.

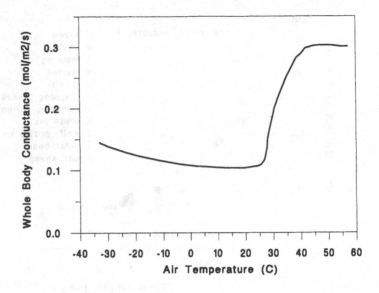

FIGURE 12.5. Variation of tissue and coat conductance of white crown sparrows with temperature (data from Mahoney and King, 1977).

Clearly this range of conductance is not achieved simply by vasodilation and vasoconstriction. Webster et al. (1985) showed that in mourning doves posture, ptiloerection, and other responses to cold could substantially alter conductance.

It is interesting that conductance increases from its minimum with both increasing and decreasing temperature. The increase with increasing temperature is for thermoregulation, but the increase as temperatures drop below freezing is probably the result of shunting blood to appendages at these cold temperatures to avoid freezing them.

Wind has a major effect on the thermal resistance of clothing and animal coats. Campbell et al. (1980) analyzed much of the available data on windspeed dependence of coat conductance and obtained the equation:

$$g(u) = g(0)(1 + cu) \tag{12.18}$$

where $g(u)$ is the conductance (for heat or vapor) in wind, $g(0)$ is the conductance of the coat at zero windspeed, and c is a constant that depends on the wind permeability of the coat. Campbell et al. (1980) show variation in c between 0.03 s/m and 0.23 s/m, with typical values for dense coats 3 to 4 cm thick being around 0.1 s/m. Therefore expect a 10 m/s wind to approximately double the animal conductance.

Rain can also substantially alter the conductance of animal coats. Webb and King (1984) compared the conductance of wet and dry coats under a range of conditions, and found that, on average, the conductance of wet coats is about double that of dry coats. Part of the decrease is due to decreased coat thickness, but a more important part is the latent heat transport.

12.5 Qualitative Analysis of Animal Thermal Response

Now that you have a general understanding of the behavior of the various terms in the energy balance equation, we return to Eq. (12.11) and relate it to strategies for thermoregulation in endotherms. Figure 12.6 is an idealized example of the response of an endotherm placed in a metabolic chamber as the temperature of the chamber is varied. Each of the zones in the diagram can be related to the energy budget equation for the animal (Eq. (12.11)). Within the thermoneutral zone the animal is able to balance its energy budget by adjusting body conductance, g_{Hb}. This has no direct metabolic cost, so the metabolic rate remains constant over this range. Typically the thermoneutral range includes the range of temperatures in the normal living environment of the animal and therefore changes with the season.

The lower critical temperature is at the lower end of the thermoneutral zone. At this temperature the animal has reached its minimum conductance and heat loss from further reduction in environmental temperature must be balanced by an increase in metabolic rate. As said previously, maximum aerobic rate is about ten times basal rate. A metabolic rate of 500 W m^{-2} would allow animals with even moderate insulation to survive the coldest temperatures on earth. In practice it is unusual to find metabolic costs for thermoregulation above about 150 W m^{-2}, about three times the basal rate.

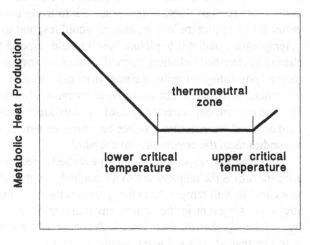

FIGURE 12.6. Thermoneutral diagram for an endotherm.

At the top of the thermoneutral zone the upper critical temperature marks the point at which whole body conductance is maximum. As the upper critical temperature is approached latent heat loss is increased by some energy-requiring process such as panting or sweating. (Sweating is not a passive process; sweat has about half the concentration of salts as normal body fluids so salts must be removed before secreting the sweat.) Since evaporative cooling requires some energy, the metabolic rate must increase somewhat as is shown in the diagram. As environmental temperature increases, body temperature also increases somewhat and this increases metabolic rate. At these high temperatures the only avenue available for balancing the energy budget is the latent heat term. When air temperature is equal to body temperature, the latent heat loss must equal the metabolic heat production. If the air temperature is higher than body temperature, this additional heat gain must also be dissipated as latent heat. The rate of latent heat loss depends heavily on the vapor pressure of the environment, and possibly on the boundary layer conductance, but maximum values of latent heat loss for fairly arid environments range from 200 to 400 W m^{-2}.

The strategy for balancing the energy budget over a wide range of environmental conditions should now be clear. The preferred mode is by varying conductance. Below the thermoneutral zone, metabolic energy is required to balance the energy budget, while above the thermoneutral zone evaporation of water is required.

12.6 Operative Temperature

We return now to the definition of the operative temperature T_e that was introduced in Eq. (12.11). There we explained that the operative temperature combines air temperature and radiation in a single equivalent temperature. This is a convenient way to represent the animal environment for at least two reasons. First, temperature is intuitively useful because it is easier for us to picture how an animal would respond to a 20°C change in temperature than it is to picture how it would respond to a 400 W m^{-2} change in absorbed radiation. Second, a lot of the knowledge we have on thermal physiology of animals comes from experiments conducted in environmental chambers. The operative temperature allows us to use results of these experiments directly in outdoor situations where the combined radiation and temperature produce an operative temperature equivalent to conditions in the environmental chamber.

The operative temperature (sometimes called the equivalent blackbody temperature) is the temperature of a blackbody cavity (with air temperature equal to wall temperature) that provides the same heat load (or cold stress) as is present in the natural environment of the animal. Another way of saying that the heat load is the same in the two environments is to say that $M - \lambda E$ for the animal is the same in the two environments. Therefore a mathematical definition of the operative temperature can be obtained by substituting Eq. (12.11) for $M - \lambda E$ in Eq. (12.10) and solving for T_e. The result is:

$$T_e = T_a + \frac{R_{abs} - \varepsilon_s \sigma T_a^4}{c_p(g_r + g_{Ha})}. \tag{12.19}$$

The operative temperature is the air temperature plus or minus some temperature increment which depends on absorbed radiation, wind speed, characteristic dimension of the animal, and temperature. In a blackbody cavity (a room or metabolic chamber) where the wall and air temperatures are equal, the absorbed radiation is equal to the emittance at air temperature, so $T_e = T_a$. Outdoors, in the shade or under clouds, and with high wind (high conductance) the radiation increment is small. Under a clear night sky R_{abs} is smaller than the emittance at air temperature, so $T_e < T_a$. In bright sunshine, the operative temperature can be much larger than air temperature.

Example 12.2. As an example of the calculation of T_e, we find the operative temperature for a person in a 1 m/s wind, 30° C air temperature, and full sun. We assume the environmental conditions are those given in Example 11.4. If we assume θ, the angle between the solar beam and the axis of the person, is 60°, then $F_p = 0.26$ (Fig. 11.6). For dark clothing, we assume $\alpha_s = 0.8$ and $\varepsilon_s = 0.97$, so (refer to Example 11.4 for details):

$$R_{abs} = 0.8 \times (0.26 \times 1160 + 0.5 \times 86 + 0.5 \times 273)$$

$$+ \, 0.97 \times (0.5 \times 407 + 0.5 \times 496) = 823 \, \frac{W}{m^2}.$$

An average characteristic dimension for a person (legs, arms, body, etc.) is $d = 0.17$ m, so the boundary layer conductance (forced convection with naturally turbulent wind) is:

$$g_{Ha} = 1.4 \times 0.135 \times \sqrt{\frac{1 \, m/s}{0.17 \, m}} = 0.46 \, \frac{mol}{m^2 s}.$$

The radiative conductance is:

$$g_r = \frac{4\sigma T_a^3}{c_p} = \frac{4 \times 5.67 \times 10^{-8} \frac{J}{m^2 s \, K^4} \times 303^3 K^3}{29.3 \frac{J}{mol \, K}} = 0.22 \, \frac{mol}{m^2 s}.$$

The operative temperature is (Eq. (12.19)):

$$T_e = 30C + \frac{823 \, \frac{W}{m^2} - 464 \, \frac{W}{m^2}}{29.3 \, \frac{J}{molC} \left(0.46 \, \frac{mol}{m^2 s} + 0.22 \, \frac{mol}{m^2 s} \right)} = 48° \, C.$$

When we want to make the point that a day is extremely hot we often say something like "it was a hundred degrees [F] in the shade," implying that one would feel much hotter than 100 degrees in the sun. The operative temperature conveys this same sentiment quantitatively. It adds a temperature increment to the air temperature to indicate the temperature of a room which would feel the same as the heat load in the sun. In the example just given, the equivalent temperature would be uncomfortably hot (well above body temperature) even though the air temperature is not uncomfortably high.

If windspeed in this example were increased to 3 m/s, T_e would decrease to 43° C. If white clothing with $a_s = 0.3$ were worn, T_e would be 36° C at $u = 1$ m/s.

12.7 Applications of the Energy Budget Equation

We return briefly to the animal energy budget equation (Eq. (12.11)) to consider some applications. The metabolic rate, latent heat loss, body temperature, and body conductance are primarily physiological, and have upper and lower bounds set by the physiological makeup of the animal. By determining the limits of these variables, the extremes of environment (T_e) can be predicted that can be tolerated by the animal. The combination of minimum body temperature, maximum sustainable metabolic rate, minimum conductance, and minimum latent heat loss defines the lower lethal limit for the animal. The combination of maximum allowable body temperature, minimum metabolic rate, maximum conductance, and maximum latent heat loss defines the upper lethal limit. The animal cannot survive extended periods of time in environments below its lower lethal limit or above its upper lethal limit. These limits define a kind of climate space in which the animal can reside. The climate space is a function of both air temperature and absorbed radiation.

In addition to being useful for predicting animal behavior, the energy budget equations can be used to predict the food energy required to maintain a favorable body temperature. If the operative temperature of the environment is specified, and the body temperature and conductance are known, the energy budget equation can be used to compute the metabolic rate needed to balance the energy budget. This is just the metabolic requirement for thermoregulation, but other energy sinks are usually small compared to the requirement for thermoregulation.

Example 12.3. How much food is required for thermoregulation by a 1.5 kg rabbit in an environment with $T_e = 0°C$? Assume $d = 0.1$ m and $u = 1$ m/s.

Solution. Since T_e is given, there is no need to consider the radiative environment of the rabbit, but when considering both day and night conditions in a typical rabbit environment T_e and T_a are likely to be about the same. Equation (12.11) can be used to find M. We need to know the body temperature, the heat and vapor conductances, and the latent heat loss. We assume body temperature is 37° C, and that the combined respiratory and skin latent heat loss is 20 percent of the metabolic rate. There are three conductances for heat loss, the convective-radiative g_{Hr}, the coat g_{Hc}, and the tissue g_{Ht}. The convective-radiative conductance is the sum of the forced convection conductance in Table 7.6 and the radiative conductance:

$$g_{Hr} = 1.4 \times 0.135 \sqrt{\frac{1 \text{ m/s}}{0.1 \text{ m}}} + \frac{4 \times 5.67 \times 10^{-8} \frac{J}{m^2 s \, K^4} \times 273^3 K^3}{29.3 \frac{J}{mol \, K}}$$

$$= 0.77 \frac{mol}{m^2 s}.$$

An estimate of the coat conductance can be obtained from Fig. 12.4 for rabbit. A mean conductance is around $0.045 \text{ mol m}^{-2} \text{ s}^{-1}$. Correcting this for wind effects using Eq. (12.18) gives: $g(1) = 0.045 \times (1 + 0.1 \times 1) = 0.05 \text{ mol m}^{-2} \text{ s}^{-2}$. There are no data for rabbit tissue conductance in Table 12.2, but the numbers shown there suggest it might be around $0.5 \text{ mol m}^{-2} \text{ s}^{-1}$. Substituting these values into Eq. (12.11) gives:

$$g_{Hb} = \frac{0.05 \times 0.5}{0.05 + 0.5} = 0.046 \frac{mol}{m^2 s}$$

$$M(1 - 0.2) = \frac{29.3 \frac{J}{mol \, C} \times 0.77 \frac{mol}{m^2 s} (37 \text{ C} - 0 \text{ C})}{1 + \frac{0.77 \text{ mol } m^{-2}s^{-1}}{0.046 \text{ mol } m^{-2}s^{-1}}}$$

so

$$M = 59 \frac{W}{m^2}$$

From Eq. (12.13), the area of a 1.5 kg animal is 0.13 m^2, so the energy requirement of the rabbit is

$$0.13 \text{ m}^2 \times 59 \frac{W}{m^2} = 7.6 \text{ W}.$$

The caloric content of glucose as 15.7 MJ/kg, so a kilogram of glucose would last

$$\frac{15.7 \times 10^6 \text{ J}}{7.6 \frac{J}{s}} = 2.05 \times 10^6 s \quad \text{or} \quad 24 \text{ days}.$$

A kilogram of dry grass would last less than half that long because of inefficiencies in absorption in the digestion process. Efficiency factors are known for many animals and diets.

12.8 The Transient State

Short periods of intense activity, such as running, flying, climbing, or exposure to high winds at cold temperatures are common in animals. During these times the heat storage term q in Eq. (12.1) is not zero, and there is a positive or negative storage of heat in the body. Transient state energy budgets are important to the animal, but have not been examined in much detail by researchers.

If we assume that the thermal conductivity and heat capacity of an animal's core are large compared to that of the coat and peripheral tissues, we can write:

$$q = \frac{V}{A} \rho_b c_b \frac{dT_b}{dt} \tag{12.20}$$

where V is the body volume, A is surface area, ρ_b and c_b are the body density and specific heat, and T_b is body temperature. Equation (12.11) can be rewritten to include transient effects as:

$$M - \lambda E - c_p g_H (T_b - T_e) = \frac{V}{A} \rho_b c_b \frac{dT_b}{dt}. \tag{12.21}$$

Equation 12.21 is difficult to integrate for homeotherms at temperatures within their control band because M, λE, and r_{Hb} are functions of T_b. For poikilotherms, we can assume $M - \lambda E = 0$; then integration of Eq. (12.21), taking all but T_b as constant, gives:

$$\frac{T_{b2} - T_e}{T_{b1} - T_e} = \exp(-t/\tau) \tag{12.22}$$

with:

$$\tau = \frac{\rho_b c_b V}{c_p g_H A}. \tag{12.23}$$

Equation (12.22) can be used to find the time required for a poikilotherm to change from T_{b1} to T_{b2} when T_e is changed. The symbol τ represents the time constant of the animal. It has units of seconds, and is an index of response time. At $t = \tau$ the system will have changed to within $1/e = 0.37$ of the total change from T_{b1}, the initial temperature, to T_{b2}, the final temperature. Thus animals with large time constants could survive exposure to environments outside their climate space for relatively long times. From Eq. (12.23) it can be noted that large volume, small surface area, and small thermal conductance maximize τ. Thus animals that weigh only a few grams respond quickly to environmental changes and are close to equilibrium temperature. Animals weighing many kilograms would seldom be at steady state and could survive short-term extremes in exposure much more readily.

A transient analysis can also be used to indicate thermal behavior of homeotherms when they are subjected to environments that are sufficiently harsh to force complete commitment for their temperature control systems. Under these conditions M, λE, and g_{Hb} become constant at their maximum or minimum values and it is possible to integrate Eq. (12.21). The resulting equation is similar to Eq. (12.22) except that the final temperature, rather than being T_e, is $T_e + (M - \lambda E)/g_H c_p$ because of metabolic heat and evaporation.

12.9 Complexities of Animal Energetics

The models we have presented for organism–environment interaction can be very useful for analyzing organism response to environment and understanding the most important factors in the animal environment. There

are many cases, however, where our simplifying assumptions are too restrictive, and can lead to incorrect conclusions. The limitations we have
imposed on latent heat loss exclude any analysis of sweating. A more
complete analysis, however, will be given in Ch. 13. We also failed to
consider heat loss by conduction to the ground or other substrate (even
though the equations for that are given in Ch. 8). Perhaps the most serious
omissions are a failure to consider the possibility that radiation can penetrate the animal coat, and the failure to consider the three-dimensional
nature of the animal. To add these complexities goes beyond the objectives of this book, but excellent work has been done in both areas, and we
briefly refer to the results of that work.

Our energy balance equations are essentially for a one-dimensional animal. We assume that the heat is well enough mixed internally to maintain
an essentially constant internal temperature. We also chose a single characteristic dimension and a single R_{abs} value for the animal in spite of the
fact that we know both of these values vary widely over the surface of the
animal. Coat conductance also varies substantially from place to place depending on the thickness of the coat and exposure to wind. Bakken (1981)
addressed these issues with what he calls a two-dimensional operative
temperature model. This new model just divides the animal up into many
zones (head, legs, body in sun, body in shade, etc.), each of which can be
adequately analyzed by an equation similar to Eq. (12.11). An operative
temperature for each zone is also computed. The overall energy budget is
then just the area-weighted average of all zones. From this kind of analysis
he concludes that in strong wind or sun the one-dimensional model can
give substantially different results than the two-dimensional model. In
one example, the operative temperature from the two-dimensional model
was 6°C lower than for the one-dimensional model.

If radiation penetrates the coat of an animal, the location of energy
absorption ceases to be the outer boundary of the coat. Dissipation of
heat, however, still occurs at the outer boundary, so the effective radiation
heat load on the animal is higher. This is a kind of miniature greenhouse
effect. Walsberg et al. (1978) determined that, for small animals with
high boundary layer conductances, radiation penetration is important in
determining the optimum coat color for animals in desert environments.
Solar radiation penetrates to deeper depths in white coats than black.
Even though the total energy absorbed by a black coat is much greater
than that absorbed by a white one, the additional heat load from radiation
penetration of the white makes the black coat more suitable for desert
environments. Observations of coat color in desert dwelling animals seem
to confirm this result.

In sparse animal coats, both long and shortwave radiation penetrate the
coat, and it becomes impossible to treat the animal–environment interface
as a definite boundary as we have in this chapter. To deal with it properly
as a continuum, computer models must be used. Porter et al. (1994) have
developed such models and have shown them to work well in ecological
applications. The model has the advantage that it properly treats all of the

complexities of the animal–environment interaction. The disadvantage is that it provides little opportunity for understanding the physical principles involved in the exchange processes except to the person who creates the model.

12.10 Animals and Water

The water budget for an animal can be written in the same form as the energy budget, namely: water in − water out = stored water. Unlike the energy budget, the water budget can seldom be analyzed as a collection of steady-state processes. Intake is from free-water sources, metabolism, and water in the animal's food. Water is taken in discrete events, not continuously. As we have shown, cutaneous and respiratory water loss are relatively continuous, and determined largely by animal activity and environment. Water is also lost in feces and urine, though this is often only a small fraction of the total water loss. Water loss reduces the amount of water stored in the blood, tissues, or digestive tract, and may decrease the osmotic potential (increase concentration) of body fluids.

MacMillian and Christopher (1975) had used urinary osmotic potential of kangaroo rats (*Dipodomys merriami*) in the desert as an index of water balance. Their data (Fig. 12.7) show only slight variations in plasma osmotic potential through the season. Urine osmotic potential decreased in the summer and increased in the winter, indicating higher water deficits in the summer. Urine osmotic potentials of some other desert rodent species were less well correlated with seasonal temperature changes

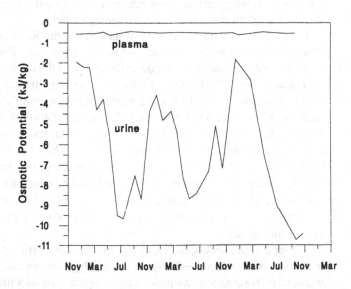

FIGURE 12.7. Osmotic potentials of plasma and urine of desert dwelling *Dipodomys merriami* over a three-year period (data from MacMillen and Christopher, 1975).

because of changes in diet with season. For comparison, the osmotic potential of human plasma is -0.75 kJ/kg and human urine is generally between -2.1 and -3.3 kJ/kg.

MacMillian also conducted laboratory studies on several desert rodent species to determine their ability to maintain water balance on a diet of dry birdseed. The kangaroo rat neither lost nor gained weight, but some other species fell far short of maintaining positive water balance. Others actually gained weight. The pocket mouse (*Perognathus longimembris*) seemed to be able to maintain a particularly favorable water balance on this diet. One wonders first how these animals can get enough water from dry seeds to supply their needs, and second why the kangaroo rat would have a less favorable water balance than the smaller mouse. Some light can be shed on these questions by a simple analysis.

When an animal oxidizes food to produce heat, water is also produced. One kilogram of glucose, when oxidized, produces 600 g of water. The ratio of latent heat from respiratory water to metabolic heat produced is $\lambda E_{pr}/M = 0.1$. We have already shown the respiratory latent heat loss for the kangaroo rat at 20°C to be $0.02M$. The skin latent heat loss (Eq. (12.16)) is 5 W/m^2, if we assume $C_{vs} - C_{va} = 40$ mmol/mol and $g_{vs} = 2.8$ mmol m^{-2} s^{-1} (Table 7.2). Equation (12.11) can now be used to find M at $T_e = T_a = 20$°C. We assume $g_{Hb} = 0.14$ mol m^{-2} s^{-1} for the kangaroo rat and 0.21 mol m^{-2} s^{-1} for the pocket mouse (estimates from Figure 12.4 and Table 12.2). Also, assume that $g_{Hr} = 0.8$ mol m^{-2} s^{-1}. The metabolic rates at 20°C, from these assumptions are 88 W/m^2 for the mouse and 65 W/m^2 for the rat. The ratio of water produced to water evaporated is $E_{pr}/E_{ev} = 0.1M/(5 + 0.02M)$. For the rat, the ratio is 1.03, and for the mouse, 1.3. These calculations are crude, but they show that the animals produce enough metabolic water to supply their water requirements without any additional water input. They also show that the more favorable water balance of the mouse is the result of the higher metabolic rate it requires to maintain constant body temperature. This metabolic requirement increases as temperature decreases and is apparently too high during the winter months for the mice to remain active because they hibernate during the winter.

References

Bakken, G. S. (1981) A two-dimensional operative-temperature model for thermal energy management by animals. J. Therm. Biol. 23-30.

Bernstein, M.H. (1971) Cutaneous water loss in small birds. Condor 73:468-469.

Calder, W. A. and J. R. King . (1974). Thermal and caloric relations of birds. p. 259-413 in D. S. Farner and J. R. King, eds. Avian Biology, V. 4. New York: Academic Press.

Campbell, G. S., A. J. McArthur, and J. L. Monteith . (1980). Windspeed dependence of heat and mass transfer through coats and clothing. Boundary Layer Meteorol. 18:485-493.

Cena, K. and J. L. Monteith (1975) Transfer processes in animal coats, II. Conduction and convection. Proc, R. Soc. Lond. B. 188:395-411.

Kerslake, D. McK. (1972) The Stress of Hot Environments. London: Cambridge University Press.

Lasiewski, R. C., M. H. Bernstein, and R. D. Ohmart (1971) Cutaneous water loss in the Roadrunner and Poor-will. Condor 73:470-472.

MacMillian, R. E. and E. A. Christopher (1975) The water relations of two populations of noncaptive desert rodents. Environmental Physiology of Desert Organisms. (N.F. Hadley, ed.) New York: John Wiley.

Mahoney, S. A. and J. R. King (1977) The use of the equivalent blackbody temperature in the thermal energetics of small birds. J. Thermal Biology 2:115-120.

Monteith, J. L., and M. H. Unsworth (1990) Principles of Environmental Physics, 2nd ed. London, Edward Arnold.

Porter, W. P. and D. M. Gates (1969) Thermodynamic equilibrium of animals with environment. Ecol. Monogr. 39:245-270.

Porter, W. P., J. C. Munger, W. E. Stewart, S. Budaraju, and J. Jaeger (1994) Endotherm energetics: from a scalable individual-based model to ecological applications. Aust. J. Zool. 42: 125-162.

Robinson, D. E., G. S. Campbell, and J. R. King (1976) An evaluation of heat exchange in small birds. J. Comp. Physiol, 105:153-166.

Schmidt-Nielsen, K. (1969) The neglected interface: the biology of water as a liquid-gas system. Quart. J. Biophys. 2:283 -304.

Schmidt-Nielsen, K. (1972) How Animals Work. London: Cambridge University Press.

Schmidt-Nielsen, K., J. Kanwisher, R. C. Lasiewski, J. E. Cohn, and W. Le Bretz (1969) Temperature regulation and respiration in the Ostrich. Condor 71:341-352.

Scholander, P. F., V. Walters, R. Hock, and L. Irving (1950) Body insulation of some arctic and tropical mammals and birds. Bio. Bull. 99:225-236.

Walsberg, G. E., G. S. Campbell, and J. R. King. (1978) Animal coat color and radiative heat gain: a re-evaluation. J. Comp. Physiol. 126:211-222.

Webb, D. R. and J. R. King . (1984). Effects of wetting on insulation of birds and mammals coats. J. Therm. Biol.

Webster, M. D. (1985) Heat loss from avian integument: effects of posture and the plumage. Indiana Academy of Sci.94:681-686.

Problems

12.1. Compare the respiratory and skin water loss for humans and mice when air temperature is 20° C and air humidity is 0.4. Assume the expired air temperature for the mouse is the same as for the kangaroo rat in Fig. 12.3.

12.2. Estimate the upper and lower lethal limit environments for a sparrow. For this, assume that the maximum sustainable metabolic rate for thermoregulation is $3M_b$, and the maximum latent heat loss is 150 W m^{-2}.

12.3. How much food does a 100 kg caribou need (kg/day) to survive an arctic winter with average T_e of -20°C? Assume $u = 3$ m/s.

12.4. What is the operative temperature for a sunbather standing on a beach at noon on a clear day, with $T_a = 30$°C, $u = 2$ m/s, and $\alpha_s = 0.7$?

Humans and their Environment

13

Human–environment interaction involves the same principles discussed in Ch. 12. However, we need to look at three additional factors. They are clothing, sweating, and comfort. These are examined by considering survival in cold environments, survival in hot environments, and the human thermoneutral energy budget. The variables that need to be considered are metabolic rate, surface area, latent heat exchange, body temperature, and body (clothing and tissue) conductance.

13.1 Area, Metabolic Rate, and Evaporation

The total body area in square meters (often called the DuBois area in honor of DuBois and DuBois (1915) who first proposed the formula) can be calculated from:

$$A = 0.2 m^{0.425} h^{0.725} \tag{13.1}$$

where m is the body mass in kilograms and h is height in meters. As a rough rule of thumb, body area of adults can be estimated from:

$$A = 0.026 \, m. \tag{13.2}$$

Metabolic rates can be calculated using Eq. (12.14), but a better guide can be obtained from measurements. Table 13.1 gives values of M for various activity levels. These activity levels conform quite well to our rules of thumb of $M_b = 30 - 50 \text{ W/m}^2$ and

$$M_{\max} = 10 M_b.$$

The published values for caloric content of foods are normally in units of kilocalories (called calories in the food literature) rather than joules. Assuming a 2 m^2 surface area for a person, W m^{-2} in Table 13.1 can be converted to kcal of food intake per hour. The conversion factor is:

$$\frac{\text{kcal}}{\text{hr}} = x \frac{\text{J}}{m^2 s} \times 2m^2 \times \frac{1 \text{ cal}}{4.18 \text{ J}} \times \frac{1 \text{ kcal}}{1000 \text{ cal}} \times \frac{3600 \text{ s}}{\text{hr}} = 1.7x.$$

Therefore, according to the numbers in Table 13.1, desk work would consume 160 kcal/hr and sleeping 85 kcal/hr. For eight hours of sleep

TABLE 13.1.   Rates of metabolic heat production for humans

Activity	M (W/m²)
Sleeping	50
Awake, resting	60
Standing	90
Working at a desk or driving	95
Standing–light work	120
Level walking at 4 km/hr or moderate work	180
Level walking at 5.5 km/hr or moderately hard work	250
Level walking at 5.5 km/hr with a 20-kg pack or sustained hard work	350
Short spurts of very heavy activity such as in climbing or sports	600

Data from Landsberg (1969)

and 16 hours standing, the daily caloric requirement would be around 3100 kcal. If a person performed hard physical labor for 12 hr/day and rested for the remaining 12 hours, the caloric intake would need to increase to 6000 kcal/day. For those who exercise for weight control, one hour of strenuous exercise is worth about 600 kcal in excess food intake. The caloric content of fat is 40 kJ/g, so strenuous exercise for 1 hr would use 63 g of fat. One might conclude that regulation of caloric intake is an easier mode of weight control that exercise. As a note of caution, remember that the values in Table 13.1 are for thermoneutral temperatures. If additional metabolic energy is required for thermoregulation (Eq. (12.11)) this must be added to the values in Table 13.1.

Latent heat is lost through respiration and through water loss directly from the skin. In Ch. 12 we derive an expression for respiratory latent heat loss, and find it to be around 0.1 M in relatively dry environments (Eq. (12.15)). In more moist environments, it is smaller. Evaporation from the skin in the absence of thermal sweating is called insensible perspiration, and can be calculated from Eq. (12.16) using the appropriate value for skin conductance from Table 12.1. Under typical conditions ($e_{va} = 1.0$ kPa, $p_a = 101$ kPa, and $g_{vs} = 5.4$ mmol m^{-2} s^{-1}), $\lambda E_s = 12$ W/m². This is a little over twice the respiratory latent heat loss at $M = M_b$.

The core temperature of the body depends mainly on metabolic heat production until environmental conditions become too severe for thermoregulation. A convenient equation expressing the relationship between metabolic rate and core temperature is (Kerslake, 1972):

$$T_b = 36.5 + 4.3 \times 10^{-3} M \qquad (13.3)$$

where M is in W/m².

Resistance to heat transfer in the human body is, as with other homeotherms, subject to vasomotor control. The tissue conductance (g_{Ht}) varies, within limits, to balance the energy budget. The limits given in Table 12.2 are $g_{Ht} = 0.46$ mol m^{-2} s^{-1} for vasoconstriction and

2.8 mol m^{-2} s^{-1} for vasodilation. These values were calculated from Kerslake (1972, Fig. 7.22). Monteith gives a range of 0.35 to 1.4 mol m^{-2} s^{-1}. The difference is probably due to acclimatization of subjects or possible subject-to-subject variation. In any case, we use the range 0.46 to 2.8 mol m^{-2} s^{-1} for our calculations.

Clothing conductance for humans is more difficult to treat than coat conductance for animals because of the extremely wide possible range of clothing available (down parkas to bathing suits). Normal indoor clothing has a conductance of around 0.4 mol m^{-2} s^{-1} in still air. In moving air, this is drastically increased, as common experience will verify. In the absence of conductance measurements for a given assemblage of clothing, one can use estimates based on windspeed, permeability, thickness, and ventilation of the clothing.

13.2 Survival in Cold Environments

Equation (12.11) will be used as the basis for our examination of energy and thermal resistance requirements for humans. Consider first the lowest temperature at which a human can survive. This can be found by assuming extreme values for M, g_{Hb}, λE, and g_{Ha}. If we assume $d = 0.17$ m, $u = 3$ m/s, $\lambda E_r = 0.1M$, $\lambda E_s = 12$ W/m^2, and $T_b = 36°$ C then the lowest equivalent temperature for survival can be calculated for various resistances and metabolic rates. From Table A.3, with $T_a = 0°$ C, $g_r = 0.16$ mol m^{-2} s^{-1}. The boundary layer conductance is:

$$g_{Ha} = 1.4 \times 0.135\sqrt{\frac{3 \text{ m/s}}{0.17 \text{ m}}} = 0.79 \frac{\text{mol}}{\text{m}^2\text{s}}.$$

The convective-radiative conductance $g_{Hr} = 0.16 + 0.79 = 0.95$ mol m^{-2} s^{-1}. These values are substituted into Eq. (12.11), along with the body temperature and latent heat loss, and the equation is solved for operative temperature to give:

$$\begin{aligned}
T_e &= T_b - \frac{(M - \lambda E_r - \lambda E_s)(g_{Hb} + g_{Hr})}{c_p g_{Hb} g_{Hr}} \\
&= 36 - \frac{(0.9M - 12)(g_{Hb} + 0.95)}{29.3 \times 0.95 g_{Hb}}.
\end{aligned} \tag{13.4}$$

This equation ignores a small temperature dependence of the radiative conductance and the metabolic rate and also assumes that skin latent heat loss is independent of temperature. It does, however, show the main effects of T_e and g_{Hb} on M. These are shown in Fig. 13.1 where M is plotted as a function of T_e for three values of conductance.

The highest value of M is for no clothing, the second is for a conductance comparable to a heavy wool business suit, and the third is equivalent to a good quality winter sleeping bag. It can be seen that survival is possible at quite low temperatures, even without clothing, if metabolic rate can be kept high.

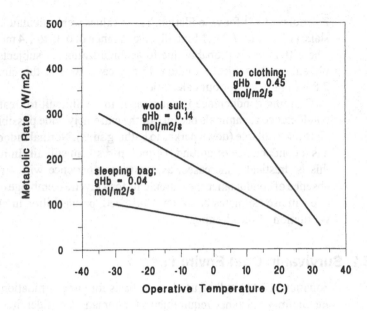

FIGURE 13.1. Metabolic rate required to balance the energy budget at various operative temperatures, for three values of g_{Hb}.

Darwin (1832) made some interesting observations on survival among the natives of Tierra del Fuego under conditions which must have been near the limits of survival. He reports:

The climate is certainly wretched: the summer solstice was now passed, yet every day snow fell on the hills, and in the valleys there was rain accompanied by sleet. The thermometer generally stood about 45F, but in the night fell to 38 or 40. . . . while going one day on shore near Wallaston Island, we pulled alongside a canoe with six Fuegians. These were the most abject and miserable creatures I anywhere beheld. On the east coast the natives, as we have seen, have guanaco cloaks, and on the west, they possess sealskins. Amongst these central tribes the men generally have an otter skin, or some small scrap about as large as a pocket handkerchief, which is barely sufficient to cover their backs as low down as their loins. It is laced across the breast by strings, and according as the wind blows, it is shifted from side to side. But the Fuegians in the canoe were quite naked, and even one full-grown woman was absolutely so. It was raining heavily, and the fresh water, together with the spray, trickled down her body. In another harbor not far distant, a woman, who was suckling a recently-born child, came one day alongside the vessel, and remained there out of mere curiosity, whilst the sleet fell and thawed on her naked bosom, and on the skin of her naked baby! . . . At night, five or six human beings, naked and scarcely protected from the wind and rain of this tempestuous climate, sleep on the wet ground coiled up like animals.

Perhaps a more useful way to plot Eq. (13.4) is to show the conductance required for different levels of activity and operative temperature. Using this graph the clothing thermal conductance required for any given activity

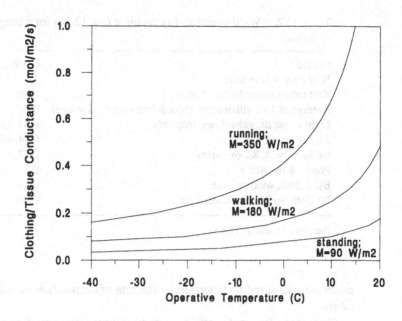

FIGURE 13.2. Thermal conductance required for survival in cold at various operative temperatures and activity levels.

and environment could be found. This is shown in Fig. 13.2. The following example illustrates the use of Fig. 13.2.

Example 13.1. Assume you are going outdoors when $T_e = -20°$ C. Find the coat plus tissue conductance needed for standing, walking, and running.

Solution. The required conductances can be read directly from Fig. 13.2. If you intend standing for long periods of time at $M = 90$ W/m², you would need $g_{Hb} = 0.05$ mol m⁻² s⁻¹. For walking you would need 0.1 mol m⁻² s⁻¹. And for running you would need 0.25 mol m⁻² s⁻¹.

13.3 Wind Chill and Standard Operative Temperature

Figure 13.2 shows the required conductance for the prevailing temperature and wind conditions. Wind has a small effect on this required conductance. However the effect of wind on clothing conductance can be large and must be taken into account when choosing the amount of clothing necessary to provide the required conductance.

Equation (12.18) gives the wind speed dependence of animal coat conductance and also works well for clothing conductance. The permeability factor, c in the equation, however, depends strongly on the fabric from which the clothing is made. Table 13.2 shows c values for a range of

TABLE 13.2.   Wind permeability factor, c (eq. 12.18) for a range
of fabrics

Fabric	c (s/m)
Very open weave shirt	1.1
Knit cotton undershirt or T shirt	0.86
Average of 13 civilian shirts (broadcloth or oxford weave)	0.61
Light worsteds, gabardines, tropicals	0.44-0.51
Seersucker suiting	0.44-0.48
Uniform twill, 8.2 oz. Army	0.27
Poplin, 6 oz. Army	0.2
Byrd cloth, wind resistant	0.16
JO cloth, special wind resistant	0.10

Data from Newburgh (1968)

materials, measured by determining the rate of evaporation through the
fabric.

Since these data are for effects of wind on vapor transport, they are not
ideal for computing effects of wind on heat transport, but lacking more
direct information we use these values for both heat and vapor. According
to these figures, a 10 m/s wind would double the conductance of JO cloth,
and a 1 m/s wind would double the conductance of a very open weave
shirt.

The effect of wind on clothing and boundary conductance is addressed
by another thermal index, the standard operative temperature, T_{es}. The
standard operative temperature, like the operative temperature, combines
several environmental variables into a single environmental index which
has dimensions of temperature. The operative temperature combined radiation and air temperature into a single equivalent temperature. The
standard operative temperature adds wind effects. Standard operative
temperature is the temperature of a uniform enclosure with still air which
would result in the same heat loss from an animal or person as occurs in
the windy, outdoor condition under investigation. The popular term for
T_{es} is the wind chill factor.

To derive an equation for T_{es}, start with the energy budget equation
(Eq. (12.11)). By definition, $M - \lambda E$ is the same for the person in the
standard enclosure and the person in the natural environment. Therefore
the following can be written:

$$M - \lambda E = \frac{c_p g_{Hr}(T_b - T_e)}{1 + g_{Hr}/g_{Hb}} = \frac{c_p g_{Hrs}(T_b - T_{es})}{1 + g_{Hrs}/g_{Hbs}}$$

where the s subscripts on the operative temperature and the conductances
indicate the standard (still air enclosure) conditions. Solving for T_{es} gives:

$$T_{es} = T_b - \frac{(g_{Hbs} + g_{Hrs})g_{Hb}}{(g_{Hb} + g_{Hr})g_{Hbs}}(T_b - T_e). \qquad (13.5)$$

There are two important things to note about Eq. (13.5). First, when the conductances are equal to the standard values, then $T_{es} = T_e$, so the standard operative temperature and the operative temperature are the same. The second is that $T_{es} = T_e = T_b$ for an ectotherm which does not control body temperature by internal heat production. People sometimes assume that, since wind makes us cold, it also makes plants, snakes, and spiders cold. It can be seen from Eq. (13.5) that this is not true (except to the extent that wind reduces T_e).

Equation (13.5) can be used to derive a wind chill chart similar to those used by the weather service. To simplify this we assume that the clothing conductance is low enough so that the boundary layer, tissue, and radiative conductances can be ignored. We also assume that the wind dependence of clothing conductance is given by Eq. (12.18). Equation (13.5) then becomes:

$$T_{es} = T_b - (1 + cu)(T_b - T_e). \qquad (13.6)$$

Comparison of this equation with the wind chill chart given by Landsberg (1969) indicates that the value of c used for wind speeds below about 10 m/s is 0.046. This is lower than even the most wind resistant fabrics given in Table 13.2, so it apparently assumes a best case. Wind chill with more permeable clothing would be more serious than the standard chart indicates. Landsberg's chart shows little change in wind chill at wind speeds above 10 m/s. This may be the result of his basing the wind chill relation on boundary-layer conductance, which increases in proportion to the square root of wind speed, rather than on the permeability of clothing, which increases more nearly linearly with wind speed. Figure 13.3 shows wind chill temperature (standard operative temperature) as a function of operative temperature (near air temperature for these wind conditions) for three values of wind speed.

To use Fig. 13.3, enter the chart at the air temperature, go to the wind speed, and read off the wind chill temperature. For example, if the air temperature were 0° C, and the wind speed were 10 m/s, then the wind chill temperature would be -17° C. This would mean that even though the air temperature is only 0° C, the outdoors would feel as cold to you as a room with still air would at -17° C.

13.4 Survival in Hot Environments

The same considerations apply to survival in hot environments as do for determining the upper lethal limit of animals. However, one additional factor needs considered—that of sweating. The rate of sweat evaporation may be either environmentally or physiologically controlled. If the skin surface is wet, the rate of water loss from sweating is given by Eq. (12.16) with $g_{vs} \to \infty$. If the skin surface is not wet, latent heat loss is controlled by sweat rate. Control of sweat rate is still not entirely understood, but apparently it involves sensing of surface heat flux (Kerslake 1972). Thus, changes in metabolic rate or external environment can cause changes in

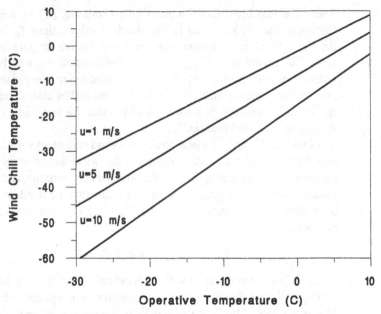

FIGURE 13.3. Wind chill chart. Lines are from eq. 13.6 with $c = 0.046$.

sweat rate. This seems reasonable, since the requirement for body temperature control is that the heat budget be balanced. Maintaining a balanced heat budget as metabolic rate or equivalent temperature increase, without increasing body temperature, requires increased heat dissipation. The human body appears to have a hierarchy of physiological responses in order to increase dissipation. As heat load increases, tissue thermal conductance increases first, then sweating begins. Finally, body temperature begins to rise. All of these responses are accompanied by increases in skin temperature. Attempts have been made to relate the first two responses to skin temperature, with only partial success. Body temperature begins to increase when skin temperature reaches about 36° C. One becomes extremely uncomfortable at skin temperatures of 36° C or greater.

Since we are primarily interested in the highest heat loads that can be sustained by humans at steady state, we need only consider the maximum rate of sweating that can be sustained. Maximum sweat rate is extremely variable, depending on acclimatization of subjects and duration of exposure. Rates as high as 4 kg/hr have been reported for short time periods (Newburgh, 1968). Typical rates are much lower than this, and a value commonly used for heat stress calculations is around 1 kg/hr. The latent heat flux equivalent to this is around 380 W/m², which we take as the maximum rate of sweating for an average person. For a heat-stressed person the latent heat loss will therefore be either 380 W/m² (physiologically limited, skin remains dry) or the value predicted by Eq. (12.16) with nonlimiting skin conductance (environment limited, skin remains wet), whichever is smaller. Figure 13.4 gives λE_s as a function of the combined

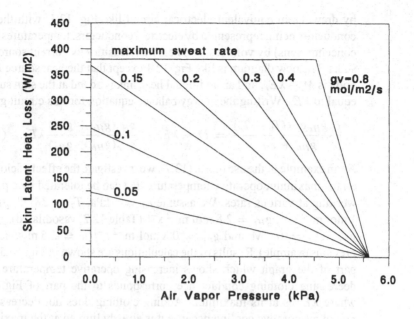

FIGURE 13.4. Latent heat loss from the skin of heat-stressed humans as a function of vapor pressure and total vapor conductance. Skin temperature is assumed to be 35 C.

clothing and boundary layer conductance, and atmospheric vapor pressure. Boundary layer conductance for vapor transport is given by Eq. 7.33. In a 2.5 m/s wind, boundary layer conductance,

$$g_{va} = 1.4 \times 0.147 \sqrt{\frac{2.5 \text{ m/s}}{0.17 \text{ m}}} = 0.8 \frac{\text{mol}}{\text{m}^2\text{s}}.$$

The rightmost line in Fig. 13.4 therefore corresponds to the evaporation rate from wet skin without clothing.

Example 3.2. What clothing conductance would allow skin to remain dry in a heat-stressed person if the vapor pressure of the air is 1.5 kPa?

Solution. Consulting Fig. 13.4, with $e_a = 1.5$ kPa, it is found that a conductance of 0.2 mol m^{-2} s^{-1} limits latent heat loss, but a conductance of 0.3 does not. A conductance between these two, say 0.25 mol m^{-2} s^{-1}, should therefore allow the skin to remain dry. This is the combined clothing and boundary layer conductance. If the boundary layer conductance is 0.8 mol m^{-2} s^{-1}, then the clothing conductance would be

$$g_{vc} = \frac{1}{\frac{1}{0.25} - \frac{1}{0.8}} = 0.36 \frac{\text{mol}}{\text{m}^2\text{s}}.$$

Survival under heat-stress conditions can be predicted using the energy budget equation, but it needs rederived without the assumption that λE_s is combined with λE_r, as it is in Eq. (12.11). This is most easily done by drawing an equivalent electrical circuit like Fig. 12.1, with thermal conductors being represented by electrical conductors, temperatures (heat concentrations) by voltages, and heat flux densities by current sources or sinks. The new diagram is like Fig. 12.1 except that the heat source in the body is $M - \lambda E_r$, and an additional heat sink is added at the skin surface equal to λE_s. Writing the energy balance equation for this circuit gives

$$\frac{c_p g_{Hr} g_{Hb}(T_b - T_e)}{g_{Hb} + g_{Hr}} = M - \lambda E_r - \frac{g_{Hb}(g_{Hc} + g_{Hr})}{g_{Hc}(g_{Hb} + g_{Hr})} \lambda E_s. \quad (13.7)$$

As an example of the use of Eq. (13.7), we investigate the effect of clothing on the maximum operative temperature that can be tolerated by a person working at various rates. We assume $e_a = 1$ kPa, $T_b = 38°$ C, $g_{Hr} = 1$ mol m^{-2} s^{-1}, $g_{Ht} = 2.8$ mol m^{-2} s^{-1} (Table 12.2, vasodilated), $g_{vc} = g_{Hc}$, $\lambda E_r = 0.1 M$, and $g_{va} = 0.8$ mol m^{-2} s^{-1}($u = 2.5$ m/s; see the previous example). Results of the calculations are shown in Fig. 13.5. The part of the graph which shows increasing operative temperature with decreasing clothing conductance corresponds to the part of Fig. 13.4 where λE_s is at its maximum. Adding clothing does not decrease the rate of evaporative cooling because it is already limited at the maximum sweat rate of the person. The clothing does, however, decrease the heat load on the person because the environment temperature is higher than

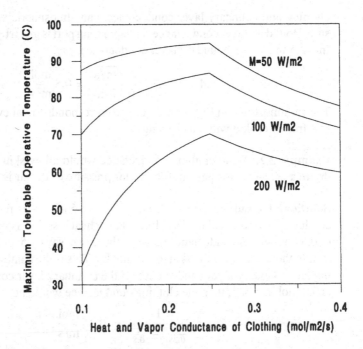

FIGURE 13.5. Maximum tolerable operative temperature for a person as a function of clothing conductance. Vapor pressure is 1 kPa, wind speed if 2.5 m/s.

the body temperature. Adding clothing therefore allows the person to tolerate a hotter environment. In a desert, with low vapor pressure and high solar loads, adding clothing (up to a point) decreases, rather than increases heat load on a person. The inflection point of the graph occurs when coat conductance becomes small enough to start controlling water loss.

Keep in mind that Fig. 13.5 is for a low vapor pressure. It does not apply at higher vapor pressure where any decrease in clothing conductance would reduce latent heat loss. If the atmospheric vapor pressure is high enough to keep the skin wet without clothing, then any addition of clothing will decrease dissipation of heat. This brings out the point that clothing must be matched to environment to be most useful in minimizing heat stress. Proper clothing for one hot environment would not necessarily be proper clothing for another.

13.5 The Humid Operative Temperature

Much effort has gone into deriving a single index that will indicate environmental heat stress for humans. The environmental variables that affect heat stress are radiation, temperature, vapor pressure, and diffusion conductances to heat and vapor. For cold stress, where latent heat loss is treated as a fixed value relatively independent of environment, the standard operative temperature adequately combined radiation and heat transfer characteristics of the environment into a single number. An appropriate energy budget equation was then used to indicate the strain imposed by a given stress, the stress being indicated by the operative temperature. It would seem reasonable to attempt to extend the operative temperature concept to include atmospheric vapor pressure. If this could be done, it would again enable the combination of all relevant environmental variables into a "stress index," and with an appropriately derived energy budget equation, could indicate the resulting strain on the individual.

The derivation proceeds in a way similar to the derivation of the operative temperature in Ch. 12: substitute Eq. (12.16) for λE_s and Eq. (12.19) for T_e into Eq. (13.7) to get an energy budget equation in terms of physiologic and environmental variables. The vapor mole fraction difference in Eq. (12.16) can be approximated using the Penman transformation (discussed in detail in Ch. 14) to give:

$$e_s - e_a = e_s(T_s) - e_s(T_a) + e_s(T_a) - e_a \cong \Delta(T_s - T_a) + D \quad (13.8)$$

where Δ is the slope of the saturation vapor pressure versus temperature function (Ch. 3) and D is the vapor deficit of the atmosphere. The slope Δ has a fairly strong temperature dependence. If its value is taken at the average of T_s and T_a then Eq. (13.8) is almost exact. Taking Δ at the average of T_b and T_a gives adequate accuracy for our purposes here. With these substitutions, the energy budget equation becomes:

$$\left(1 + \frac{s}{\gamma^*}\right)(T_b - T_a) - \frac{R_{abs} - \varepsilon_s \sigma T_a^4}{c_p g_{Hr}}$$

$$= \frac{(M - \lambda E_r)(g_{Hb} + g_{hr})}{c_p g_{Hr} g_{Hb}} - \frac{D}{\gamma^* p_a} \tag{13.9}$$

where

$$\gamma* = \frac{\gamma(1/g_{vs} + 1/g_{vc} + 1/g_{va})}{1/g_{Hc} + 1/g_{Hr}}, \quad \text{and} \quad s = \frac{\Delta}{p_a}.$$

The humid operative temperature (Gagge, 1981) is the temperature of a uniform enclosure, with a humidity of 100 percent. For a person in such a humid chamber, Eq. (13.9) reduces to:

$$\frac{\gamma*}{s + \gamma*}(M - \lambda E_r) = \frac{c_p g_{Hr} g_{Hb}(T_b - T_{eh})}{(g_{Hb} + g_{Hr})(1 + s/\gamma*)} \tag{13.10}$$

where T_{eh} is the temperature of the enclosure, or the humid operative temperature. The required definition of humid operative temperature is obtained by subtracting Eq. (13.10) from Eq. (13.9) to give:

$$T_{eh} = T_a + \frac{\gamma^*}{s + \gamma^*}\left(\frac{R_{abs} - \varepsilon_s \sigma T_a^4}{c_p g_{Hr}} - \frac{D}{\gamma^* p_a}\right). \tag{13.11}$$

These equations are more general forms of Eqs. (12.11) and (12.19) since, as g_{vs} becomes small, γ^* becomes large, making $\gamma^*/(s + \gamma^*)$ go to one and terms with γ^* in the denominator go to zero. Equation (13.11) is the heat stress index we were seeking, since it combines all of the relevant environmental variables into a single equivalent temperature. As with T_e, the operative temperature T_{eh} is equal to the body temperature of an ectotherm (with $M - \lambda E = 0$) for the environment specified by T_a, R_{ni}, D, and g_{Hr}.

The temperature of a copper sphere covered with black, moistened cloth and filled with water, has been used to determine wet globe temperature and these measurements have been related to human comfort in hot environments. The exchange properties of such a system are not identical to those for a human, so the temperature measured in this way would not be the humid operative temperature. It is possible, though, that wet globe temperature would correlate with T_{eh}.

When some air movement is present, the skin surface is wet, and clothing vapor and heat transfer resistances can be assumed equal, γ^* becomes almost equal to γ. At body temperature, the term $\gamma/(s + \gamma)$ is 0.17. The second term in Eq. (13.11) is therefore quite small and T_{eh} is usually only a few degrees different from T_a.

13.6 Comfort

In their usual activities, humans are generally not so concerned with minimum conditions for survival as they are with comfort. For this, the energy budget approach still gives the needed answers. We need only put in physiological parameters that we think represent comfort. These, of course, vary considerably from individual to individual. For our purposes we assume that a person is comfortable if $T_b = 37°$ C, and $g_{Ht} = 1$ mol m^{-2} s^{-1}. For normal indoor conditions we take $g_{Ha} = 0.2$ mol m^{-2} s^{-1} and $g_r = 0.2$ mol m^{-2} s^{-1} so $g_{Hr} = 0.4$ mol m^{-2} s^{-1}. Thermal conductance of normal indoor clothing is assumed to be 0.4 mol m^{-2} s^{-1}. Combining Eqs. (12.15) and 12.16, and assuming both skin and expired air temperature are 34° C, gives:

$$\lambda E = (1.8M + 2.38)\frac{(5.32 - e_a)}{p_a}. \qquad (13.12)$$

When this expression is substituted into Eq. (12.11) estimates of comfortable operative temperature can be obtained. These are plotted in Fig. 13.6 for vapor pressures of 0.5 and 3 kPa. If the room wall temperature is equal to air temperature then operative temperature and air temperature are equal. Figure 13.6 shows that for normal active metabolic activity ($M = 90$ W/m^2), a comfortable room temperature at low vapor pressure would be 23° C. In a humid room, the comfortable temperature would be 21° C. Thus it is possible to reduce room temperature and maintain comfort if the air is humidified. This has been suggested as a means for reducing heating costs. A more complicated analysis would be necessary to determine whether humidifying the air would actually reduce heating costs since one would need to compare the cost of evaporating the water to

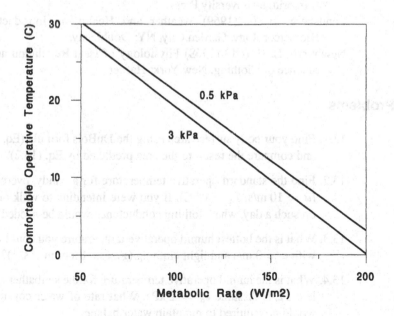

FIGURE 13.6. Comfortable operative temperature for two air vapor pressures as a function of metabolic rate.

humidify the air with the cost of keeping the room a few degrees warmer. Figure 13.6 also shows that a relatively small change in activity results in a fairly large change in comfortable temperature. This is also confirmed by common experience.

Many other aspects of comfort could be investigated using the energy budget equation. For example, one sometimes feels cold or hot in a room even when the thermometer indicates an air temperature of 22° C. This is particularly true in rooms with large windows. If we were to measure the window temperatures (and possibly wall or ceiling temperature) we would likely find that they are significantly above or below air temperature. The radiation from these cold or hot areas therefore produces an operative temperature that is quite different from air temperature. Thus we see that our radiant energy environment is very important to comfort, even indoors.

References

Darwin, Charles (1832) Journal of Researches into the Natural History and Geology of the Countries Visited During the Voyage of H.M.S. Beagle Round the World. London: John Murray.

Dubois, D. and E. F. Dubois (1915) The measurement of the surface area of Man. Arch. Intern. Med. 15:868-881.

Gagge, A. P. (1981) Rational temperature indices of thermal comfort. p.79-98 in K. Cena and J. A. Clark (eds.), Bioengineering, Thermal Physiology and Comfort. Amsterdam: Elsevier.

Kerslake, D. McK. (1972) The Stress of Hot Environments. London: Cambridge University Press.

Landsberg, H. E. (1969) Weather and Health, an Introduction to Biometeorology. Garden City, NY: Doubleday.

Newburgh, L. H. (ed.) (1968) Physiology of Heat Regulation and The Science of Clothing. New York: Hafner.

Problems

13.1. Find your body surface area using the DuBois formula (Eq. (13.1)) and compare the result to the area predicted by Eq. (13.2).

13.2. Find the standard operative temperature for a windy, overcast day ($u = 10$ m/s, $T_a = 0°$ C). If you were intending to walk outdoors on such a day, what clothing conductance would be needed?

13.3. What is the hottest humid operative temperature you could work at with $u = 3$ m/s and light clothing ($g_{Hc} = 1$ mol m^{-2} s^{-1})?

13.4. What is the humid operative temperature for the sunbather in Problem 12.4? Assume $e_a = 2$ kPa. What rate of water consumption would be required to maintain water balance?

Plants and Plant Communities 14

Our discussions in Chs. 12 and 13 focus on determining which environments were energetically acceptable to animals and on energetic costs of living in those environments. Similar questions apply to the study of plants and plant communities. In this chapter we are interested in the environmental factors that determine temperatures and transpiration rates, and in the factors that control carbon assimilation. The energy budget again plays the central role in these analyses.

While an animal can choose its environment to best suit its energetics, plants are pretty well stuck with whatever environment happens to exist at the location and time of their growth. Over generations, selection and adaptation result in leaf morphologies, canopy structures, etc. which give the plants native to a particular environment a competitive advantage for that location. Desert plants that experience frequent shortages of water, for example, tend to have narrow leaves, while leaves of plants from more moist environments may be much larger. We might ask ourselves what environmental limitations there are to leaf size and other leaf characteristics related to energy exchange, or whether there is an optimum leaf form for a particular leaf environment. Answers to questions like these have obvious application in managed ecosystems such as agriculture. The answers are likely to be found both in studying the physics of energy and mass exchange, and in observing the characteristics of natural plants and plant communities in different environments.

Three factors must be favorable for a leaf to remain alive. Average net photosynthesis must be positive and the leaf water potential and temperature must remain within nonlethal bounds. Mature leaves apparently have no mechanism for importing sugars, so a leaf which is not able to maintain a positive net photosynthesis abscises. Net photosynthetic rate is determined by environmental factors and by the water balance of the leaf. To get a clear picture of plant responses to environment the environmental effects on leaf temperature, leaf water balance, and photosynthesis need to be considered.

In this chapter we also consider these processes in plant communities, but only in a simple sense. So-called big leaf models are often used to

model temperature, transpiration, and photosynthesis in plant communities. The equations for such models are similar to those for individual
leaves, but with conductances adjusted appropriately. The big leaf models
for plant communities are presented in this chapter. In Ch. 15 we present
the more complex models that deal with plant communities as collec-
tions of individual leaves. We first consider the effects of environment
on transpiration and leaf or canopy temperature, and then present several
models that relate photosynthesis to light, temperature, and transpiration.
Finally, we combine the photosynthesis and energy balance equations to
predict response of photosynthesis to plant and environmental variables,
and attempt to specify optimum leaf form for a particular environment.

14.1 Leaf Temperature

The temperature of a leaf is determined, as with a poikilothermic animal,
by the energy budget of the leaf. In Chs. 12 and 13 equations for the temperature of poikilotherms are derived. For a dry system, where latent heat
exchange is a small and predictable fraction of the total energy budget,
the operative temperature gives the temperature of the poikilotherm. For
a wet system, where latent heat loss is an important part of the energy
budget, the humid operative temperature (Eq. (13.11)) is the temperature
of the poikilotherm. The leaf normally is a wet system, so its temperature is equal to the humid operative temperature. We derive that equation
again here to clarify its connection to the energy budget for a leaf. If heat
storage and metabolic heat production are assumed negligible, the energy
budget for a leaf is:

$$R_{\text{abs}} - L_{oe} - H - \lambda E = R_{\text{abs}} - \varepsilon_s \sigma T_L^4 - c_p g_{Ha}(T_L - T_a)$$

$$- \lambda g_v \frac{e_s(T_L) - e_a}{p_a} = 0 \qquad (14.1)$$

where R_{abs} is the absorbed radiation, L_{oe} is the emitted thermal radiation,
H is the sensible heat loss, λE is the latent heat loss, T_L is the leaf
temperature, T_a is the air temperature, and e_a is the vapor pressure of
air. The heat conductance, from Table 7.6, is $g_{Ha} = 1.4 \cdot 0.135\sqrt{u/d}$,
where u is the wind speed and d is the characteristic dimension of the leaf
(0.72 times the leaf width). For applications in outdoor environments, the
factor of 1.4 is included. The vapor conductance g_v is the average surface
and boundary conductance for the whole leaf. Care needs to be taken in
defining the vapor conductance since abaxial and adaxial conductances
are generally not equal. Assuming the boundary layer conductances are
equal for the two sides of the leaf, the appropriate vapor conductance for
Eq. (14.1) is computed from

$$g_v = \frac{0.5 g_{vs}^{ab} g_{va}}{g_{vs}^{ab} + g_{va}} + \frac{0.5 g_{vs}^{ad} g_{va}}{g_{vs}^{ad} + g_{va}} \qquad (14.2)$$

where the superscripts ab and ad refer to abaxial and adaxial surface
conductances. Table 7.2 gives some typical surface conductances for
leaves.

Some species have stomata on only one side of the leaf (usually the abaxial side). Such leaves are called hypostomatous. The conductance for these leaves is computed from just the first term on the right of Eq. (14.2), since, for zero surface conductance, the second term is zero. Leaves with stomates on both sides of the leaf are called amphistomatous. In the special case where an amphistomatous leaf has equal abaxial and adaxial conductances, the overall conductance for the leaf is equal to the conductance for either side. For simplicity, the examples in this chapter assume equal conductances on the two sides of the leaf. Leaves of this type are most common in grasses.

Equation (14.1) shows the explicit relationships for radiant emittance, sensible heat, and latent heat loss. It therefore relates leaf thermodynamic temperature explicitly to leaf properties and environmental variables. The equation for leaf temperature could be solved using mathematical procedures for nonlinear equations, but we cannot obtain an explicit solution because of the nonlinear emittance and saturation vapor pressure terms. Since an explicit form of the equation is useful for our analyses, we obtain an approximate solution using the linearization techniques introduced in Chs. 12 and 13. First, the thermal emittance term can be linearized using Eq. (12.6) to obtain:

$$\varepsilon_s \sigma \mathbf{T}_L^4 \cong \varepsilon_s \sigma \mathbf{T}_a^4 + c_p g_r (T_L - T_a) \tag{14.3}$$

where g_r is the radiative conductance. The latent heat term can also be linearized using Eq. (13.8):

$$\lambda g_v \frac{e_s(T_L) - e_a}{p_a} = \lambda g_v \frac{e_s(T_L) - e_s(T_a)}{p_a} + \lambda g_v \frac{e_s(T_a) - e_a}{p_a}$$

$$\simeq \lambda g_v s(T_L - T_a) + \lambda g_v \frac{D}{p_a} \tag{14.4}$$

where D is the vapor deficit of the atmosphere, $s = \Delta/p_a$, and $\Delta = de_s(T)/dT$. This linearization was first used by Penman (1948) to derive the famous Penman equation for evapotranspiration. Using Eqs. (14.3) and (14.4), Eq. (14.1) can now be written as:

$$R_{abs} - \varepsilon_s \sigma \mathbf{T}_a^4 - \lambda g_v D/p_a - (c_p g_{Hr} + \lambda s g_v)(T_L - T_a) = 0. \tag{14.5}$$

The convective-radiative conductance $g_{Hr} = g_{Ha} + g_r$ has been used here. Equation (14.5) can now be readily solved for leaf temperature to obtain:

$$T_L = T_a + \frac{R_{abs} - \varepsilon_s \sigma \mathbf{T}_a^4 - \lambda g_v D/p_a}{c_p g_{Hr} + \lambda s g_v}$$

$$= T_a + \frac{\gamma^*}{s + \gamma^*} \left[\frac{R_{abs} - \varepsilon_s \sigma \mathbf{T}_a^4}{g_{Hr} c_p} - \frac{D}{p_a \gamma^*} \right]. \tag{14.6}$$

The second form is the same as Eq. (13.11) for the humid operative temperature, where $\gamma^* = \gamma g_{Hr}/g_v$. Either equation provides a straightforward way to determine leaf temperature from air temperature, radiation, wind, and vapor deficit.

Example 14.1. Find which is cooler in a hot, dry environment, a wide leaf or a narrow leaf, if both have access to an unlimited water supply. Assume leaf widths of 3 mm and 3 cm; $T_a = 38°$ C, $e_a = 1.1$ kPa, $u = 1.5$ m/s, and $R_{abs} = 750$ W/m^2. Pressure is 100 kPa. Also assume the leaves are hypostomatous with an adaxial conductunce of zero and an abaxial conductance of 0.5 mol m^{-2} s^{-1}.

Solution. From Table A.3, it can be seen that $e_s(T_a) = 6.63$ kPa, $\Delta = 0.359$ kPa, $B = 532$ W/m^2, and $g_r = 0.23$ mol m^{-2} s^{-1}. Using these values the following can be computed

$$R_{ni} = 750 - 0.97 \times 532 = 234 \text{ W/m}^2$$

$$D = 6.63 - 1.1 = 5.53 \text{ kPa}$$

$$s = 0.359 \text{ kPa C}^{-1}/100 \text{ kPa} = 0.00359 \text{ C}^{-1}$$

we also have (Table A.1), $c_p = 29.3$ J mol^{-1} C^{-1}, and (Ch. 3) $\gamma = 6.66 \times 10^{-4}$ C^{-1}. For the leaf specific calculations we have the following table.

---

10 cm wide leaf

$d = 0.7 \times 0.1 = 0.07$ m

$g_{Ha} = 1.4 \times 0.135\sqrt{\frac{1.5}{0.07}} = 0.875$ mol m^{-2} s^{-1}

$g_{va} = 1.4 \times 0.147\sqrt{\frac{1.5}{0.07}} = 0.95$ mol m^{-2} s^{-1}

$g_{Hr} = 0.875 + 0.23 = 1.105$ mol m^{-2} s^{-1}

$g_v = \frac{0.5 \times 0.5 \times 0.953}{0.5 + 0.953} = 0.164$ mol m^{-2} s^{-1}

$\gamma^* = 6.66 \times 10^{-4}\frac{1.105}{0.164} = 0.00449$ C^{-1}

$T_L = 38 + \frac{.00449}{.00449 + .00359}\left(\frac{234}{1.105 \times 29.3} - \frac{5.53}{100 \times 0.00449}\right)$
$= 35.2°$ C

---

3 mm wide leaf

$d = 0.7 \times 0.003 = 0.0021$ m

$g_{Ha} = 1.4 \times 0.135\sqrt{\frac{1.5}{0.0021}} = 5.05$ mol m^{-2} s^{-1}

$g_{va} = 1.4 \times 0.147\sqrt{\frac{1.5}{0.0021}} = 5.50$ mol m^{-2} s^{-1}

$g_{Hr} = 5.05 + 0.23 = 5.73$ mol m^{-2} s^{-1}

$g_v = \frac{0.5 \times 0.5 \times 5.50}{0.5 + 5.50} = 0.229$ mol m^{-2} s^{-1}

$\gamma^* = 6.66 \times 10^{-4}\frac{5.73}{0.229} = 0.0167$ C^{-1}

$T_L = 38 + \frac{.0167}{.0167 + .00359}\left(\frac{234}{5.73 \times 29.3} - \frac{5.53}{100 \times 0.0167}\right)$
$= 36.4°$ C

While the temperatures of the two leaves do not differ by much, it is significant and interesting that the large leaf is cooler than the small one. The greater rate of convective energy exchange associated with the smaller leaf causes the leaf to be nearer to air temperature than the larger leaf and thus warmer. This is the result of the smaller γ^* for the large leaf. In descriptive terms, this is like the sweating animal with body temperature below air temperature. When the leaf is cooler than the air, it is taking up heat from the air. It can decrease the amount of heat taken from the air, and therefore decrease its temperature, if the boundary layer conductance is low.

Equation (14.6) can be used directly to answer a rather significant ecological question. Plant ecologists commonly take air temperature as the environmental variable characterizing a site. The implicit assumption is that air temperature and plant temperature are equal, or at least related by some constant, since it is really plant temperature that determines productivity. Equation (14.6) can be used to check this assumption.

Figure 14.1 shows $T_L - T_a$ as a function of leaf characteristic dimension, stomatal conductance, and isothermal net radiation for $T_a = 30°$ C, $h_r = 0.2$, and $u = 1$ m/s. The two radiation levels correspond roughly to full sun and dark. The two stomatal conductances are roughly the highest and lowest for leaves from Table 7.2. It can be seen that small leaves remain within a few degrees of air temperature, no matter what the stomatal conductance. Large leaves have much higher leaf temperatures than small leaves when stomata are closed and lower leaf temperatures when stomata are open. However, for a wide range of leaf sizes, when stomata are open, leaf temperatures tend to remain near air temperature no matter what the leaf size. It is interesting that, for leaves with open stomata and high radiation loads, an intermediate size around 10 mm appears to give lowest leaf temperatures.

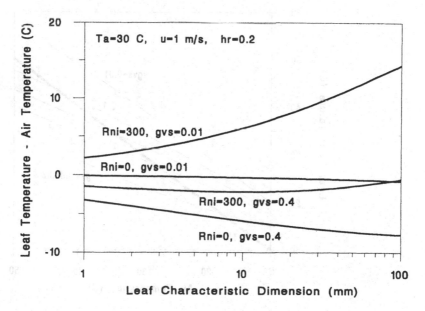

FIGURE 14.1. Difference between leaf and air temperature for various leaf dimensions, stomatal conductances, and radiation loads.

In the plant kingdom many examples of adaptations are seen which appear to be for the purpose of controlling leaf temperature near optimum levels. For example, in deserts, where water is scarce and air temperatures tend to be above optimum temperature for photosynthesis, leaves tend to be small, allowing leaf temperatures to be as near to air temperature as possible without evaporating large amounts of water. In alpine regions, where air temperature is likely to be below optimum temperature, leaves are not necessarily small, but plants tend to grow in dense clumps or cushions near the ground. This provides a large effective characteristic dimension and low wind speed, and thus results in temperatures sometimes 10 or 20° C above air temperature. Perhaps the most interesting are some species which orient their leaves or have leaf hairs which minimize the radiation load on the leaves. Ehleringer and Mooney (1978) describe a desert shrub, *Encilia farinosa*, that has relatively large leaves with high reflectance. The shrub grows along drainage channels where water is relatively more plentiful and can therefore maintain high transpiration rates. Its leaves are often several degrees below air temperature.

Equation (14.6) can be used to investigate another interesting phenomenon. A number of researchers have observed a kind of homeothermy in leaves. When air temperature is below optimum leaf temperature, leaves tend to be above air temperature, but when air temperature is above optimum temperature, leaf temperature is below air temperature. Is it possible that leaves are homeotherms? Figure 14.2 shows leaf temperatures in full sun, computed using Eq. (14.6), as a function of air temperature

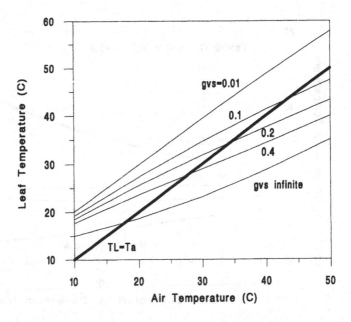

FIGURE 14.2. Leaf temperature versus air temperature for several stomatal conductances when $R_{ni} = 300$ W/m^2 and dew point temperature is 10° C.

for several values of stomatal conductance. Dew point temperature, and therefore atmospheric vapor pressure, is kept constant as air temperature changes (as it is during the day). The heavy black line shows the slope which would result if a one degree increase in air temperature resulted in a one degree increase in leaf temperature. It can be seen that, in fact, temperatures of transpiring leaves do tend to remain more constant than the air temperature. It is obviously not, however, an active response of leaves to maintain constant temperature, since a wet wash rag (infinite surface conductance) maintains the most constant temperature of all. The apparent homeothermy is just a normal response of an evaporating surface. Obviously, this does not prevent it from being beneficial to the plant.

14.2 Aerodynamic Temperature of Plant Canopies

An equation for predicting the aerodynamic temperature of plant canopies can be derived just as was done for a leaf. This temperature is referred to as aerodynamic temperature because it is derived from the solution of aerodynamic transport equations. There are two differences: the energy budget of a plant canopy must include the heat storage in the soil, and the boundary layer conductances and absorbed radiation must be computed using the appropriate equations from Chs. 7 and 11. The energy budget equation is:

$$R_{\text{abs}} - L_{oe} - G - H - \lambda E = 0 \qquad (14.7)$$

where G is the flux density of heat into or out of the soil. Using the same substitutions used for leaves, the canopy temperature equation is obtained:

$$
\begin{aligned}
T_{\text{canopy}} &= T_a + \frac{R_{\text{abs}} - \varepsilon_s \sigma T_a^4 - G - \lambda g_v D / p_a}{c_p g_{Hr} + \lambda s g_v} \\
&= T_a + \frac{\gamma^*}{s + \gamma^*} \left[\frac{R_{\text{abs}} - \varepsilon_s \sigma T_a^4 - G}{c_p g_{Hr}} - \frac{D}{\gamma^* p_a} \right].
\end{aligned}
\qquad (14.8)
$$

The boundary layer conductances for heat and vapor are computed from (Table 7.6):

$$
g_{Ha} = g_{va} = \frac{0.4^2 \hat{\rho} u(z)}{\left[\ln\left(\frac{z-d}{z_M}\right) + \Psi_M \right] \left[\ln\left(\frac{z-d}{z_H}\right) + \Psi_H \right]}.
\qquad (14.9)
$$

The total vapor conductance for the canopy is $g_v = 1/(1/g_{vc} + 1/g_{va})$. The canopy conductance g_{vc} is the weighted sum of stomatal conductances of all leaves in the canopy, plus a conductance for evaporation from the soil. Kelliher et al. (1994) showed that maximum canopy conductances tend to be conservative and fairly independent of leaf area index. They say that maximum g_{vc} is around three times maximum g_{vs}.

14.3 Radiometric Temperature of Plant Canopies

Vegetative canopies are exceedingly complex, being composed of many leaves, branches, stems, soil, etc. Even though the aerodynamic temperature can be defined by Eq. (14.8), its relation to true thermodynamic temperature is virtually impossible to discover. However, another canopy temperature is easily measured. This is the radiometric temperature. The radiometric temperature of a blackbody (unity emissivity) surface (T_{BB}) is estimated from a direct measurement of thermal radiant flux density ($\Phi(T_{BB})$) by inverting an integral of Eq. (10.4) over the wavelength band of sensitivity of the infrared radiometer. If a surface is not a blackbody, then adjustments must be made for emissivity. Norman and Becker (1995) discuss radiometric temperature and thermal emissivity in detail. Infrared radiometers that are used to measure radiometric temperature are called infrared thermometers. Because infrared thermometers are intended to estimate the temperature of a surface and be minimally influenced by the intervening atmosphere, usually they are sensitive only to wavelengths where the atmosphere is relatively transparent (between 8 and 13 μm wavelengths, see Fig. 10.6). From satellites, atmospheric influences of 3 to 10° C are not uncommon even in the most transparent wavelength bands. We know that the integral of Eq. (10.4) over all wavelengths is equal to σT_{BB}^4. If we assume the radiant flux density in the 8 to 13 μm wavelength band is proportional to $\mathbf{T}^4$ (a good approximation, but not perfect) we can work with $\mathbf{T}^4$ instead of complicated functions of the blackbody integral. Unfortunately the thermal emissivity of natural surfaces between 8 and 13 μm may not be equal to the broad-band (4 to 80 μm) thermal emissivity (particularly for soils), and the 8 to 13 μm emissivity must be known to obtain radiometric temperatures. Fortunately most full-cover vegetative canopies have thermal emissivities in the 8 to 13 μm wavelength band of 0.98 to 0.99. The reason for this is discussed near the end of Ch. 15.

Even for a blackbody, the radiometric temperature, the thermodynamic temperature, and the aerodynamic temperature resulting from a surface energy balance (Eq. (14.8)) will all be equal only if the surface and its surroundings are in thermodynamic equilibrium (they have a constant, uniform temperature). Since this rarely occurs in nature, in general these temperatures are not expected to be interchangeable. The aerodynamic temperature depends on the areodynamic conductance between the atmosphere and various parts of the surface that are at different temperatures. The radiometric temperature depends on the fourth power weighting of the absolute temperature of the parts of the surface that make up the view of the infrared thermometer. Because radiometric temperarure can depend on radiometer view angle and aerodynamic temperature does not, the two will generally be different. Consider a partial-cover canopy with hot dry soil (50° C) and cool transpiring leaves (25° C), a common occurance. If an infrared thermometer pointed at this surface from directly overhead views 40 percent soil and 60 percent vegetation, assuming emissivity to be 1.0, the radiometric temperature would be 35.7° C $((0.6 * 298^4 + 0.4 * 323^4)^{1/4} = 308.7$ K). If the view zenith angle of the infrared thermometer were changed to 85°, the view would be mainly vegetation and the indicated temperature would change to about 25° C. Since the aerodynamic temperature would have to be the same for the two infrared thermometer view angles, clearly the two temperatures can be quite different.

Since leaf and canopy temperature are determined, in part, by stomatal conductance, and conductance is determined, in part by availability of water to the plant, an effort has been made to sense plant water stress from airplanes or satellites using thermal imaging of vegetation temperature. For a dense, full-cover canopy, radiometric temperature may approximate aerodynamic temperature within 1° C. However, Eq. (14.8) provides a means of finding canopy conductance from measurements of canopy and air temperature only if wind, radiation, and vapor deficit are known. Without measuring these confounding variables, or at least making the measurements during times when they are relatively constant (such as midday on clear days with high vapor deficits), determining water stress very accurately using this technique is difficult. Even when water stress can be determined from a canopy temperature measurement, additional information is needed to determine whether a crop needs irrigation. Stomata may close and canopy temperature increase for many reasons, only one of which is a soil water deficit.

14.4 Transpiration and the Leaf Energy Budget

Another useful application of the leaf energy budget is to compute the transpiration rate. Written in terms of latent heat loss (useful in the energy budget) the transpiration rate for a leaf can be computed using Eq. (6.7):

$$\lambda E = \lambda g_v \frac{e_s(T_L) - e_a}{p_a} \tag{14.10}$$

where E is the vapor flux density (mol m^{-2} s^{-1}), g_v is the vapor conductance (mol m^{-2} s^{-1}), λ is the latent heat of vaporization for water (44 kJ/mol), and $e_s(T_L)$ and e_a are the vapor pressures at the leaf surface and in the air. This form of the equation is not very useful because the vapor pressure in the leaf depends on leaf temperature, and the leaf temperature is not usually known. However, Eq. (14.10) can be combined with the energy budget equation to obtain an equation for the transpiration rate which is independent of leaf temperature. To do this first use the Penman transform (Eq. (14.4)) to separate Eq. (14.10) into two parts, one dependent on the temperature difference between leaf and air, and the other dependent on the vapor deficit of the air. Then substitute Eq. (14.6) for $T_L - T_a$ to eliminate leaf temperature from the equation. With some manipulation this results in:

$$\lambda E_{\text{leaf}} = \frac{s(R_{\text{abs}} - \varepsilon_s \sigma T_a^4) + \gamma^* \lambda g_v D / p_a}{s + \gamma^*} \qquad (14.11)$$

where $\gamma^* = c_p g_{Hr} / \lambda g_v$, that is, γ^* is a psychrometer constant, but now with the heat and vapor conductances included. If the conductances for heat and vapor are equal, then $\gamma^* = \gamma$, the thermodynamic psychrometer constant.

Example 14.2. Compare the transpiration rates of the leaves in Example 14.1.

Solution. All of the information needed to solve Eq. (14.11) was obtained in Example 14.1 except for the latent heat of vaporazation. From Table A.2, $\lambda = 43500$ J/mol. For the large leaf the latent heat loss is

$$\lambda E = \frac{0.00359 \times 234 + 0.00449 \times 43500 \times 0.164 \times 5.53/100}{0.00359 + 0.00449}$$

$$= 323 \text{ W/m}^2,$$

For the small leaf it is

$$\lambda E = \frac{0.00359 \times 234 + 0.0167 \times 43500 \times 0.229 \times 5.53/100}{0.00359 + 0.0167}$$

$$= 495 \text{ W/m}^2.$$

If mass or molar water loss values are needed, these numbers could be converted to moles per square meter per second by dividing by the latent heat of vaporization. For comparison purposes, though, the latent heat loss values are adequate. It can be seen that it takes about 50 percent more water (per unit area) to keep the small leaf cool as it does the large one, even though the small leaf is not staying as cool as the large one. If the leaf is to remain below air temperature in a hot, dry climate, there are clear advantages to large leaf size, both in terms of leaf temperature and in terms of water loss. If the leaf temperature is above air temperature (closed stomates), then the smallest leaves remain the coolest. Thus in arid climates, plants tend to have small leaves and low stomatal conductances to simultaneously conserve water and maintain leaf temperature as near air temperature as possible.

Equation (14.11) shows that the latent heat loss from a leaf is the weighted sum of two terms, the isothermal net radiation ($R_{\text{abs}} - \varepsilon_s \sigma T_a^4$) and the isothermal latent heat loss ($\lambda g_v D / p_a$). The weighting factors are $s/(s + \gamma^*)$ and $\gamma^*/(s + \gamma^*)$. As temperature increases, s increases rapidly (see Table A.3) so the higher the air temperature, the more dominant radiant energy input is in determining evaporation.

Even though Eq. (14.11) looks simple, it is not easy to guess how it will behave in all cases. For example, one might look at Eq. (14.10) and predict that increasing wind speed would increase evaporation of water from a leaf. This, however, does not take into account the effect of the

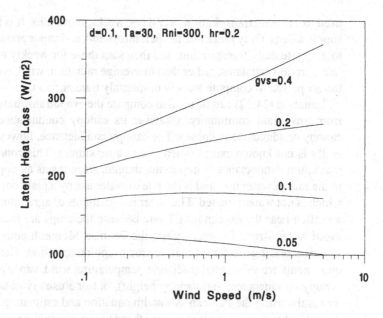

FIGURE 14.3. Latent heat loss from a leaf as a function of wind speed showing that transpiration can increase or decrease with wind speed depending on other environmental conditions.

wind on leaf temperature. Using Eq. (14.11), it is possible to include the effect of wind on boundary layer conductance for both heat and vapor. Figure 14.3 shows the evaporation rate, computed using Eq. (14.11), for different stomatal conductances. It can be seen that, with a high radiation load, increasing wind can either increase or decrease the evaporation rate. At high stomatal conductance, increasing boundary layer conductance increases transpiration rate, but at low stomatal conductance the increase in wind speed cools the leaf enough so that the decrease in vapor pressure more than compensates for the increase in boundary layer conductance and transpiration rate decreases with wind speed.

14.5 Canopy Transpiration

The equation for canopy transpiration is, again, similar to the one for leaf transpiration. We treat the canopy as a big leaf, so all that is needed is to add soil heat flux to Eq. (14.11) and compute the R_{ni} and conductances using the appropriate equations. The canopy transpiration equation is:

$$\lambda E_{\text{canopy}} = \frac{s(R_{\text{abs}} - \varepsilon_s \sigma T_a^4 - G) + \gamma^* \lambda g_v D / p_a}{s + \gamma^*}. \quad (14.12)$$

This is the well known and widely used Penman–Monteith equation (Monteith, 1965) for estimating evapotranspiration from plant communities. As we have presented it here, it appears just to provide canopy transpiration estimates at a particular instant, but it is now commonly used to estimate transpiration over days, weeks, or months. It is better to supply at least daily radiation, temperature, wind, and vapor pressure data to compute daily transpiration, and then sum these for weekly or longer transpiration estimates, rather than to average radiation, wind, etc. over a longer period to compute weekly or monthly transpiration.

Equation (14.12) can be used to compute the evapotranspiration (ET) from any plant community, whatever its canopy conductance, if the canopy conductance is known. The canopy conductance, however, generally is not known except when it is at a maximum. This condition of maximum conductance is important, though, since it sets an upper limit to the rate of water use, and is the rate of water use by a plant community which is not water stressed. The water requirements of agricultural crops are often near the maximum ET rate because the crops are managed to avoid water stress. In recent years the Penman-Monteith equation has found increasing use for estimating crop evapotranspiration. The data requirements are substantial (radiation, temperature, wind, vapor pressure, canopy conductance, and canopy height), but one usually obtains better results using the Penman–Monteith equation and estimating missing data, than by using a simpler equation that does not use such a mechanistic approach.

A particular use of the Penman–Monteith equation is for the computation of reference ET. Reference ET is the ET from a 12 cm high grass crop that completely covers the ground and is not short of water (Allen et al., 1994). The canopy conductance of the grass crop is assumed to be 0.6 mol m^{-2} s^{-1}. Equation (14.9) is used to compute boundary layer conductance. For a fixed measurement height, the conductance is just a constant multiplied by the wind speed (a number of comparisons of estimates with and without the stability parameters have shown no advantage to using them in the calculation). Allen et al. (1994) recommend using $g_{Ha} = g_{va} = 0.2u$ when wind is measured at a height of 2 m. The vapor conductance is therefore

$$g_v = \frac{0.6 \times 0.2u}{0.6 + 0.2u} \text{ mol m}^{-2}\text{ s}^{-1},$$

and the apparent psychrometer constant is

$$\gamma^* = 6.67 \times 10^{-4} \frac{0.2u}{\frac{0.12u}{0.6+0.2u}} = 6.67 \times 10^{-4}\left(1 + \frac{u}{3}\right)C^{-1}.$$

Only the convective conductance (rather than g_{Hr}) is used in computing γ^*. It is not clear why, but the estimated reference ET comes closer to the measured ET without including the radiative conductance (Allen et al., 1994). With these substitutions, computation of reference ET is straightforward. The ET of tall crops, or crops that do not completely cover the ground, is determined by multiplying the reference ET by an empirically determined crop coefficient.

14.6 Photosynthesis

The assimilation of carbon by leaves follows the general reaction

$$CO_2 + H_2O + light \rightarrow CH_2O + O_2$$

where CH_2O is intended to represent carbohydrate such as sucrose or starch. Assimilation involves many chemical reactions which occur inside the chloroplasts in leaf mesophyll cells and are catalyzed by numerous enzymes. The substrates for assimilation are CO_2, water, and light. The carbon dioxide comes from the atmosphere and diffuses into the leaf through the stomata. The water is available in excess within the leaf, since the biochemical reactions occur within the highly hydrated cell. Light is from the sun and is the photosynthetically active radiation (PAR) discussed in Ch. 10. Besides carbohydrate, oxygen is an important byproduct of photosynthesis.

The leaf environment supplies the CO_2 and the light for photosynthesis and controls the temperature of the leaf. The enzymes that catalyze the photosynthetic reactions are all strongly temperature dependent, so leaf temperature can play an important role in determining the assimilation rate for a leaf. We are primarily interested in knowing how assimilation responds to environment, but this requires some understanding of the biochemistry, since the biochemistry and the environment interact so strongly in determining how much assimilation can occur.

Most plant species fall into one of two major groupings with respect to carbon assimilation. In the most common group the primary product of photosynthesis is a three carbon sugar, so these species are called C3. A less common photosynthetic mechanism is present in tropical grasses such as maize and sugar cane. In these, the first product of photosynthesis is a four carbon compound. These species are therefore called C4. Carbon dioxide and oxygen compete for the same enzyme in C3 species resulting in the loss of some of the CO_2 in a process called photorespiration. The fixing of CO_2 into the four carbon compound in C4 species concentrates the carbon dioxide and minimizes photorespiration. The concentration of CO_2 inside the stomata of leaves is therefore much lower in C4 than C3 species typically resulting in higher photosynthetic rates and higher water use efficiencies.

14.7 Simple Assimilation Models

Two general approaches have been used to derive models relating assimilation to environment. One is more empirical and the other more mechanistic. Both are useful for understanding and predicting leaf–environment interaction.

The simpler model applies mainly to plant communities. Monteith (1977) observed that when biomass accumulation by a plant community is plotted as a function of the accumulated solar radiation intercepted by the community, the result was a straight line. Figure 14.4 shows

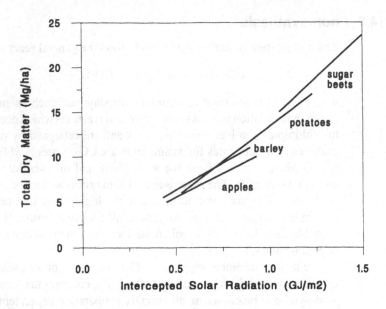

FIGURE 14.4. Total dry matter produced by a crop as a function ot total "cumulative" intercepted radiation (from Monteith, 1977).

Monteith's results. The model suggested by Fig. 14.4 is

$$A_{n,\text{canopy}} = e f_S S_t \tag{14.13}$$

where S_t is the total solar radiation incident on the canopy, f_S is the fraction of incident solar radiation intercepted by the canopy, and e is the conversion efficiency for the canopy. This conversion efficiency can be expressed in several ways; the radiation can be intercepted or absorbed as well photosynthetically active or solar radiation, and the canopy assimilation can be expressed as CO_2 or dry matter. All these possible combinations have been used and the numerical values for each is different from the others. Monteith expressed assimilation as $g\,m^{-2}\,day^{-1}$ and S_t as the total solar radiation in MJ/day, and reported e values around 1.5 g/MJ for C3 crop species.

More recently the photosynthetically active radiation has been used to estimate canopy assimilation rather than solar radiation because only the visible wavelengths are effective in photosynthesis. Furthermore, both CO_2 uptake and light can be expressed meaningfully in molar units so that the light use efficiency is dimensionless, as an efficiency should be. Sometimes the absorbed radiation is used in Eq. (14.13). Since the absorptivity of leaves is so high in the PAR band, there is little difference between absorbed and total PAR, but this is not the case for total solar radiation, since so much of the NIR is reflected. This is another reason to use PAR rather than solar. When the conversion efficiency is expressed as dry matter divided by intercepted radiation, some factors that have little to do with photosynthesis and light get included; for example, dark respiration and the composition of dry matter (fraction of carbohydrates, proteins, or lipids). The most stable conversion efficiencies are likely to be mol CO_2 (mol photons)$^{-1}$. Typical daily conversion efficiencies in these units are 0.01 to 0.03 mol CO_2 (mol photons)$^{-1}$. This is sometimes referred to as canopy light use efficiency. The conversion efficiency approach is used to estimate daily, monthly, or seasonal assimilation. One of the factors that is known to affect conversion efficiency (e) on a daily basis is the fraction of incident radiation that is diffuse versus solar beam; with diffuse radiation being more efficient.

Monteith and others have pointed out that using accumulated dry matter and intercepted radiation amounts to relating two variables that are accumulated sums. Summing any two sets of numbers, even random numbers, induces a high correlation, similar to that shown in Fig. 14.4. The fact that we get nice straight lines is therefore not necessarily an indication of a causal relationship between the two quantities. It is known from other information, though, that light and photosynthesis are causally related, so this induced correlation may add, rather than detract from the model since it makes the model very robust. The real question is whether the model is useful for prediction of dry matter production. This depends on how conservative e is. A number of experiments have shown that e is very conservative in situations where water, nutrients, and temperature do not limit plant growth. Equation (14.13) is therefore useful for predicting maximum productivity. When stresses limit growth, it is often possible to quantify their effect either in terms of a reduction in conversion efficiency, e, or a decrease in interception, f_S. This allows experiments carried out under different conditions of light availability to be compared or normalized.

The Monteith model focuses on light as the limiting substrate for photosynthesis. Another simple model can be derived by considering gas exchange. The net carbon assimilation for a leaf can be computed from:

$$A_{n,\text{leaf}} = g_c(C_{ca} - C_{ci}) \qquad (14.14)$$

where g_c is the conductance of the boundary layer and surface (stomata) for CO_2, C_{ca} is the atmospheric CO_2 concentration (around 350 μmol/mol) and C_{ci} is the CO_2 concentration in the intercellular spaces of the leaf. The subscript n on the assimilation rate means the net assimilation rate. Wong, et al. (1979) found that C_{ci} is maintained at a fairly constant value in light. Genotypes vary in the values they maintain, but the main variation is between C3 and C4 species. In C3 species values around 280 μmol/mol are common, while in C4 the values are around 130 μmol/mol. Photorespiration therefore maintains a much larger intercellular concentration in C3 leaves.

Water vapor diffuses through the same stomatal pores as CO_2, so any assimilation is accompanied by transpiration. The rate of transpiration can be computed from Eq. (14.10), (with the λs canceled). Taking the ratio of assimilation to transpiration gives:

$$\frac{A_{n,\,\text{leaf}}}{E_{\text{leaf}}} = \frac{g_c p_a (C_{ca} - C_{ci})}{g_v (e_s - e_a)}. \tag{14.15}$$

Referring to Table 7.4 it can be seen that the ratio of g_c/g_v ranges from 0.66 to 0.75 for diffusion and convection processes. Since part of the transport is by diffusion and part by convection we use a midrange value of 0.7 for the ratio. From Fig. 14.1 it can be seen that leaf temperature tends to be quite close to air temperature when stomata are open and leaves are in the sun. We could therefore approximate the vapor pressure difference between the leaf and the air by the vapor deficit of the air, D. Our simple photosynthesis model then becomes:

$$A_{n,\text{leaf}} = \frac{k E_{\text{leaf}}}{D} \tag{14.16}$$

where $k = 0.7\, p_a (C_{ca} - C_{ci})$. Tanner and Sinclair (1983) extended this model to apply to plant communities and showed that the only difference between the leaf and canopy model was the value of k used.

Relationships like Eq. (14.16) were obtained over a century ago by researchers who correlated biomass production and transpiration of crops. The fact that dry environments (with high vapor deficits) produce less biomass per unit transpiration than humid environments was also observed long before this equation was derived from gas exchange principles. The theory therefore appears to fit the observations.

Like Eq. (14.13), Eq. (14.16) applies to any leaf or canopy situation if the appropriate values for k and D are known. Equation (14.16) is more useful though if k is conservative and D is large enough so that ignoring the temperature difference between the leaves and the air does not cause too much error. Equation (14.16) is therefore not very useful under conditions of low light and high humidity. Fortunately, these are exactly the conditions for which Eq. (14.13) works well. The two equations are therefore somewhat complementary. Equation (14.16) implicitly includes light effects through the effect of radiation on E.

Equation (14.16) is useful for a number of predictions without even doing computations. For example, it predicts that dry matter production cannot occur unless there is transpiration. The amount of production which will occur per unit of water used is determined by k, which is related to the intercellular CO_2 concentration in leaves. Species with C4 metabolism maintain much lower internal CO_2 concentration than C3, so they produce more dry matter per unit water than do C3. Improvements in water use efficiency (dry matter produced per unit of water used) in a species must come mainly from decreased intercellular CO_2 concentration. This obviously has a limit and dreams of genetically engineering plants that will grow in the desert and produce dry matter without using water are obviously conjured up without much understanding of the physics of photosynthesis.

14.8 Biochemical Models for Assimilation

To investigate more detailed questions related to response of assimilation
to the leaf environment a more detailed model is needed that addresses
temperature sensitivity of enzymes and limitations by light, CO_2, and the
export of products of photosynthesis. The model we present here is from
Collatz et al. (1991). The model considers the gross assimilation rate A,
in units of $\mu mol\ m^{-2}\ s^{-1}$, to be the minimum of three potential capacities:

$$A = \min \left\{ \begin{array}{l} J_E \\ J_c \\ J_s \end{array} \right\} \tag{14.17}$$

where J_E is the light-limited assimilation rate, J_c is the Rubisco-limited
rate, and J_s is the rate imposed by sucrose synthesis.

The light-limited assimilation rate can be computed from:

$$J_E = \frac{\alpha_p e_m Q_p (C_{ci} - \Gamma^*)}{C_{ci} + 2\Gamma^*} \tag{14.18}$$

where α_p is the absorptivity of the leaf for PAR, e_m is the maximum quantum efficiency (maximum number of CO_2 molecules fixed per quantum of
radiation absorbed), Q_p is the PAR photon flux density incident on the leaf
($\mu mol\ m^{-2}\ s^{-1}$), and C_{ci} is the intercellular CO_2 concentration. The light
compensation point Γ^* is the CO_2 concentration at which assimilation is
zero. It is computed from:

$$\Gamma^* = \frac{C_{oa}}{2\tau} \tag{14.19}$$

where C_{oa} is the oxygen concentration in air (210000 $\mu mol/mol$) and
τ is a ratio describing the partitioning of RuBP to the carboxylase or
oxygenase reactions of Rubisco. In C3 species, oxygen competes with
CO_2 and τ is a measure of this competition.

The Rubisco-limited assimilation rate is computed from:

$$J_c = \frac{V_m (C_{ci} - \Gamma^*)}{C_{ci} + K_c(1 + C_{oa}/K_o)} \tag{14.20}$$

where V_m is the maximum Rubisco capacity per unit leaf area (μmol
$m^{-2}\ s^{-1}$), K_c is the Michaelis constant for CO_2 fixation, and K_o is
the Michaelis constant for oxygen inhibition. Equation (14.20) is a
rectangular hyperbola, typical of enzyme catalyzed reactions. At low
concentrations of CO_2, J_c shows a linear increase with increasing concentration, but when C_{ci} is large, J_c becomes almost constant, approaching
the value V_m. The oxygen concentration influences the initial slope of
the relationship, but not the final value reached at high CO_2 concentrations. At normal atmospheric CO_2 concentrations, decreasing the oxygen
concentration around a C3 leaf dramatically increases the photosynthetic
rate in light, but at high CO_2 levels the effect of oxygen concentration is
negligible.

The final constraint is the one imposed by the export and use of the products of photosynthesis. As in any chemical reaction, when the concentration of products builds up, the reaction slows. Sucrose synthesis is considered the most likely rate limiting step. The sucrose-limited assimilation rate is assumed, by Collatz et al. (1991) to be just

$$J_s = V_m/2. \tag{14.21}$$

Equation (14.17) implies a sharp transition from one rate limiting process to another. In reality there is a more gradual transition, with some colimitation when two rates are nearly equal. This colimitation is modeled empirically using quadratic functions. The minimum of J_E and J_c is first computed from:

$$J_p = \frac{J_E + J_c - \sqrt{(J_E + J_c)^2 - 4\theta \, J_E J_c}}{2\theta} \tag{14.22}$$

where θ represents a number between 0 and 1 that controls the abruptness of the transition from one limitation to the other. Measurements tend to give values of θ around 0.95. The second limitation is imposed by computing the minimum of J_p (from Eq. (14.22) with J_s:

$$A = \frac{J_p + J_s - \sqrt{(J_p + J_s)^2 - 4\beta \, J_p J_s}}{2\beta} \tag{14.23}$$

where β performs the same function in Eq. (14.23) that θ did in Eq. (14.22). A typical value for β is 0.98, indicating a sharp transition between J_p and J_s.

The net assimilation rate is the gross assimilation given by Eq. (14.23) minus the respiration rate for the leaf:

$$A_{n,\text{leaf}} = A - R_d. \tag{14.24}$$

Collatz, et al. (1991) compute R_d as 0.015 V_m.

The temperature response of photosynthesis is modeled by considering the temperature dependence of the model parameters. Five parameters need adjustment for temperature: K_c, τ, K_o, V_m, and R_d. The first three temperature adjustments take the same form, namely:

$$k = k_{25} \exp[q(T_L - 25)] \tag{14.25}$$

where k represents the value of any of the parameters at leaf temperature T_L, k_{25} is the value of that parameter at 25° C, and q is the temperature coefficient for that parameter. In addition to this adjustment, V_m and R_d need a high temperature cutoff. The temperature response for these parameters is:

$$V_m = \frac{V_{m,25} \exp[0.088(T_L - 25)]}{1 + \exp[0.29(T_L - 41)]} \tag{14.26}$$

and

$$R_d = \frac{R_{d,25} \exp[0.069(T_L - 25)]}{1 + \exp[1.3(T_L - 55)]}. \qquad (14.27)$$

The denominator reduces the value of $V_{m,25}$ or $R_{d,25}$ rapidly at temperatures above 41 or 55° C, respectively.

It is, of course, not practical to use this photosynthesis model for hand calculations. It can be a simple matter, though, to make a computer program which solves these equations. Table 14.1 gives values for the model parameters, as supplied by Collatz et al. (1991).

Figure 14.5 shows the temperature response predicted by the model, Fig. 14.6 shows the light response, and Fig. 14.7 shows the CO_2 response. It is important to note, in the table and in all of the figures, that our values are per unit total leaf surface area. Collatz et al. (1991) and most other researchers compute photosynthesis on a per unit projected leaf area basis. Our values are therefore half those typically found in the photosynthesis literature.

14.9 Control of Stomatal Conductance

We already mentioned the observation of Wong et al. (1979), who found that stomata tend to open or close to maintain a constant internal CO_2 concentration. Stomata must therefore be sensitive to changes in environmental CO_2 concentration, opening when it decreases and closing when it increases. Others have observed that stomata also open or close in response to the vapor deficit of the air (Lange, et al. 1971). High vapor

TABLE 14.1.   Values of parameters and constants used in the photosynthesis model. The values of b, R_d, and V_m are half those given by Collatz et al. (1991) because we assume the surface area of the leaf to be the total surface area, rather than the projected area.

Symbol	Value	Units	Temp. Coef.	Description
b	0.003	mol m^{-2} s^{-1}		intercept, B-B model
e_m	0.08	mol/mol		maximum quantum efficiency
K_c	300	μmol/mol	0.074	Michaelis constant for CO_2
K_o	300	mmol/mol	0.018	inhibition constant for O_2
m	5.6			slope parameter for B-B model
R_d	1.5	μmol m^{-2} s^{-1}	0.088	day respiration
V_m	100	μmol m^{-2} s^{-1}	0.088	Rubisco capacity
C_{ca}	340	μmol/mol		ambient CO_2 mole fraction
C_{oa}	210	mmol/mol		oxygen mole fraction
α_p	0.8			leaf absorptivity for PAR
β	0.98			colimitation factor
t	2.6	mmol/μmol	-0.056	CO_2/O_2 specificity ratio
θ	0.95			colimitation factor

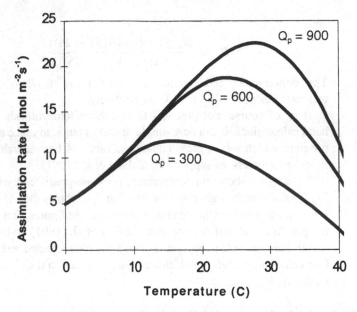

FIGURE 14.5. Temperature response of photosynthesis at three PAR levels for C_{ci} of 240 μmol mol^{-1}.

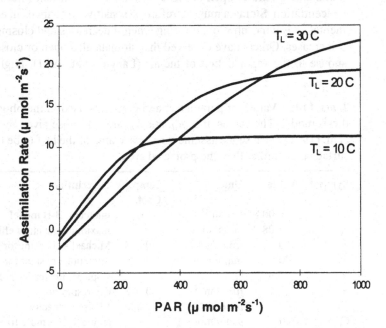

FIGURE 14.6. Photosynthesis as a function of PAR at three leaf temperatures.

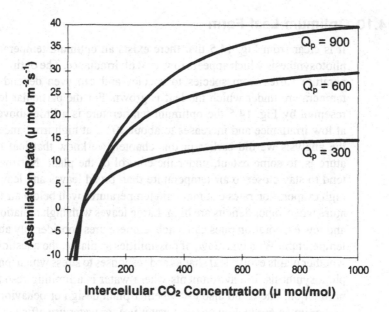

FIGURE 14.7. Photosynthesis as a function of CO_2 concentration at three light levels.

deficits (low humidities) tend to close stomata. It is not, of course, the external environment that provides the direct response, but the internal environment of the leaf. This, in turn, is controlled by external environment as well as the transpiration and assimilation rate of the leaf. Collatz et al. (1991) were able to combine all of these effects into an empirical model that looks relatively simple:

$$g_{cs} = m \frac{A_n h_s}{C_{cs}} + b \tag{14.28}$$

where h_s and C_{cs} are the humidity and CO_2 concentration at the leaf surface, A_n is the net assimilation rate, and m and b are constants determined from gas exchange studies. While Eq. (14.28) *looks* simple, it is in reality quite complex because the entire photosynthesis model determines the value of A_n; the air vapor pressure, leaf temperature, and transpiration rate determine the value of the surface humidity, and the ambient CO_2 concentration, assimilation rate, and boundary layer conductance determine C_{cs}. Since conductance determines assimilation rate, and assimilation rate determines conductance, Eqs. (14.14), (14.24), and (14.28) (with all of the equations that go into them) must be solved simultaneously to determine the assimilation of the leaf. Again, this is not something that can be done easily by hand, but can be done with a computer. The interesting thing is that, after all of the work of solving these equations, the results are almost identical to those shown in Figs. 14.5 to 14.7. Using the values in Table 14.1, the internal CO_2 concentration is controlled at 250 μmol/mol.

14.10 Optimum Leaf Form

It is clear from Fig. 14.5 that there exists an optimum temperature for photosynthesis which appears to vary with irradiance. The optimum temperature varies from species to species and can even depend on the temperature under which the leaf is grown. For the particular leaf represented by Fig. 14.5 the optimum temperature is a little above 20° C at low irradiance and increases to about 30° C at high irradiance. From the analysis we did earlier in this chapter, we know that leaf temperature is, to some extent, under the control of the plant. Narrow leaves tend to stay closer to air temperature than broad leaves and leaves with high evaporation rates can maintain temperatures well below air temperature when vapor deficits are high. Large leaves with high radiation loads and low evaporation rates can reach temperatures considerably above air temperature. With this range of possibilities available, the question arises whether plants evolve leaf shapes and responses to stress which maximize photosynthesis. In environments where water is a limiting resource for production, one could also ask whether plant design or behavior adapts to maximize production per unit water use, or water use efficiency.

We can not know whether plants maximize photosynthesis or water use efficiency, but we can investigate what size and orientation of leaves would give maximum photosynthesis or water use efficiency, and then see if leaves in that environment have that size or orientation. A few cases appear to be straightforward. The alpine cushion plants, which would be below the optimum temperature for photosynthesis most of the time if they were at air temperature, clearly benefit by radiative heating of the leaves. Their growth habit appears to be an adaptation to maximize temperature in the sun. Desert perennials, which maintain leaves throughout the summer with limited water supplies, in environments where air temperature is at or above the photosynthetic optimum, would benefit from minimizing daytime leaf temperature. Their small leaves appear to be an adaptation to keep leaves as close to air temperature as possible without evaporating large amounts of water. A number of these adaptations, and their energetic consequences, have been analyzed by Taylor (1975). At least for extreme environments, species native to those environments appear to evolve leaf shapes that tend to be optimum.

Leaf orientation is another interesting topic for investigation. Leaves of many species droop or roll when they are water stressed. This can reduce the radiation load on the leaf, decreasing leaf temperature and transpiration. On the other hand, leaves of sunflower, peanut, and many other species follow the sun, tending to increase the irradiance of the leaf. One interesting species, in this regard, is the prairie compass plant (Jurik et al., 1990). This plant grows in hot, dry environments. Leaves grow so that the flat surfaces of leaf blades face east-west to maximize radiation interception in the morning and evening and minimize it during midday. Vapor deficits are maximum during midday and leaf temperatures are higher than optimum for photosynthesis. The main assimilation times for compass grass are during morning and evening hours when vapor deficits are low, so the carbon gain per unit water loss is therefore high. Carbon gain was shown to be similar for all leaf orientations, but water use efficiency is higher for their preferred orientation.

The interactions between plants and their environment can be extremely complex. We have shown that even the behavior of the leaf temperature and transpiration models is not always intuitive or straightforward. When these are combined with the leaf photosynthesis model, and all of the interactions are in place, the result can be quite complex and difficult to predict. This appears, however, to be a fruitful area for research. The results of the work are not only useful for understanding plant adaptations to particular environments, but also to design agricultural plants for optimum production in particular environments.

References

Allen, R. G., M. Smith, A. Perrier, and L. S. Pereira (1994) An update for the definition of reference evapotranspiration. ICID Bulletin 43:1–92.

Collatz, C. J., J. T. Ball, C. Grivet, and J. A. Berry (1991) Physiological and environmental regulation of stomatal conductance, photosynthesis, and transpiration: a model that includes a laminar boundary layer. Agric. For. Meteorol. 54:107–136.

Ehleringer, J.R., and H.A. Mooney (1978) Leaf Hairs: Effects on Physiological Activity and Adaptive Value to a Desert Shrub. Oecol, 37:183–200.

Jurik, T. W., H. Zhang, and J. M. Pleasants (1990) Ecophysiological consequences of nonrandom leaf orientation in the prairie compass plant, *Silphium laciniatum*. Oecologia 82:180–186.

Kelliher, F. M., R. Leuning, M. R. Raupach, and E. D. Schulze (1994) Maximum conductances for evaporation from global vegetation types. Agric. For. Meteorol. 73:1–16.

Lange, O.L., R. Losch, E.-D. Schulze, and L. Kappen (1971) Responses of stomata to changes in humidity. Planta 100:76-86.

Monteith, J.L. (1965) Evaporation and Environment. 19th Symposia of the Society for Experimental Biology, University Press, Cambridge, 19:205–234.

Monteith, J.L. (1977) Climate and the Efficiency of crop production in Britain. Phil. Trans. R. Soc. Lond. B. 281:277–294.

Norman, J.M. and F. Becker (1995) Terminology in thermal infrared remote sensing of natural surfaces. Agric. For. Meteorol. 77:153–166.

Penman, H. L. (1948) Natural evaporation from open water, bare soil, and grass. Proc. R. Soc. A194:220.

Tanner, C. B. and T. R. Sinclair (1983) Efficient water use in crop production: research or re-search? in Limitations to Efficient Water Use in Crop Production. ASA Special Publication. American Society of Agronomy, Madison, WI.

Taylor, S. E. (1975) Optimal leaf form. Perspectives in Biophysical Ecology (D. M. Gates and R. B. Schmerl, eds.) New York: Springer-Verlag.

Taylor, S. E. and O. J. Sexton (1972) Some implications of leaf tearing in Musaceae. Ecology 53:143–149.

Wong, S.C., I.R. Cowan, and G.D. Farquhar (1979) Stomatal conductance correlates with photosynthetic capacity. Nature, 282:424–426.

Problems

14.1. When T_a is 30° C, air vapor pressure is 1 kPa, and wind speed is 2 m/s, find the temperature of a 3 cm wide amphistomatous leaf with stomatal conductance on each side of 0.3 mol m^{-2} s^{-1}, when absorbed radiation is $R_{abs} = 730$ W/m^2.

14.2. Find reference crop evapotranspiration on a clear day with total solar radiation of $S_t = 825$ W/m^2, $T_a = 30°$ C, $u = 3.5$ m/s, and $h_r = 0.3$. Assume the albedo of the reference crop is 0.2.

14.3. Use Eq. (14.16) to compare dry matter production of crops in arid and humid environments. A typical value of k for C3 crops is 0.005 kPa if assimilation is in kg dry mass per m^2 and transpiration is in kg water per m^2. Assume that the humid and arid locations both have average maximum temperatures of 27° C, but the average minimum temperatures of the humid and arid locations are 20° C and 10° C, respectively. Using the maximum and minimum temperatures, estimate the average maximum vapor deficit for each location. The D in Eq. (14.16) is the average daytime vapor deficit, which is around 0.7 times the maximum deficit. Assume 500 mm (500 kg/m^2) of water is transpired for growing the crop in each location. How much dry matter could be produced in each? For a given investment in water, is it most cost effective to irrigate crops in humid or arid environments?

14.4. How would reducing the wind speed by a factor of two affect leaf temperature compared to doubling the leaf size?

14.5. Under what conditions would you expect to find leaf temperature near to air temperature?

14.6. Under what conditions of wind, radiation, air temperature, and humidity would you expect the leaf temperature to depart most from the air temperature in a positive direction (leaf hotter than the air) and negative direction (leaf cooler than the air)?

The Light Environment of Plant Canopies

15

In Ch. 14 plant canopies are treated as big leaves. We did not worry about their structure or the details of how the leaves make up the canopy, we just assumed that we could find a canopy conductance for vapor and boundary layer conductances for heat and vapor. Combining these with the absorbed radiation and soil heat flux densities allowed us to compute canopy temperatures and transpiration rates. We even estimated carbon assimilation rates by knowing transpiration rate or light interception.

In this chapter we look in more detail at the light environment of plant canopies. Without knowing how the light is distributed on leaves within the canopy we could not use detailed photosynthesis models like the last one presented in Ch. 14 to estimate canopy photosynthesis, but a study of the light environment of plant stands is useful for many other purposes as well. In this chapter we show how to compute the fraction of radiation intercepted by a canopy and the fraction transmitted to the soil. These are important for computing assimilation using simple models like Eq. (14.13), as well as for partitioning potential evapotranspiration between evaporation (from soil) and transpiration (from leaves). We also show how to compute the change in spectral composition of light as it is transmitted and reflected by the canopy. These spectral changes have application in predicting responses of organs or organisms which are triggered by a specific ratio of red to far-red radiation and in radiometric remote sensing.

15.1 Leaf Area Index and Light Transmission Through Canopies

We use the cumulative hemi-surface area index (HSAI) L to measure the optical pathlength of radiation from the top of the canopy downward. The hemi-surface area index is one-half the surface area of leaves per unit ground area. For thin, flat leaves, the hemi-surface area index is the same as the leaf area index (LAI), which is the silhouette (one-sided) area of leaves per unit ground surface area. For more complicated shapes, like conifer needles or branches, the hemi-surface area index is not equal to the silhouette leaf area index. For example, conifer needles shaped like cylinders have a hemi-surface area index equal to $\pi/2$ times the silhouette leaf area index. Silhouette or "projected" leaf area index is not consistent in canopies with leaves that have complex shapes because the details of the projection are important and often not recorded. For example, some needles are shaped like hemicylinders and are twisted along their length a variable number of rotations, and others may have a cross section like 1/4 of a cylinder and lay flat on a planimeter with more than one orientation. Throughout this chapter we consider flat leaves and use the terms HSAI and LAI interchangeably. At the top of the canopy, $L = 0$. With increasing depth into the canopy, L increases and is equal to the total leaf area index of the canopy L_t below the canopy. For a given canopy there exists a relationship between L and the physical distance one would measure in the canopy, but the relationship is not necessarily a simple one.

Much can be learned about the role of canopy architecture in determining the relation between leaf radiative properties and canopy radiative properties by considering canopies to consist of a statistical distribution of flat leaves. You may recall from Ch. 11 that leaves typically have absorptivities of about 0.5 (Table 11.4) and canopies typically have absorptivities of about 0.8 (Table 11.2); this difference is related to the architecture of the canopy. Separating the effects of leaf spectral properties from the effect of canopy architecture is important because most leaves have similar reflectance and transmittance spectra, even though they depend on wavelength, but canopy architecture can vary widely with species, environmental condition, and time.

An idealized canopy can be constructed above some horizontal ground area of size A by randomly placing horizontal, black (leaves that do not reflect or transmit radiation) leaves each of area a above the ground. If one black leaf is randomly placed over the area A the probability that a random ray will hit this leaf is a/A. The fraction of the area A that would be in shadow by a uniform light beam at the zenith is also a/A. If this incident beam of light is thought of as being composed of a great many very small rays of light, then a fraction a/A of these tiny rays would intersect the leaf and the fraction $1 - a/A$ would pass by the leaf unintercepted. If a second leaf is placed randomly over the area A the probability that a light ray will not be intercepted by either leaf is $(1 - a/A)^2$, because the placement of the second leaf is independent of the placement of the first leaf. If N leaves are placed randomly above the area A then the probability that none of the leaves will intercept a light ray is $(1 - a/A)^N$. This describes a binomial distribution and accommodates the fact that many of the shadows of these horizontal leaves will overlap on the ground surface, A. The quantity $(1 - a/A)^N$ can be thought of as the transmittance of light from the zenith through this canopy of black, horizontal, randomly-distributed leaves. Alternatively, $(1 - a/A)^N$ can be considered to be the fraction of the ground area illuminated by the incident beam. At this point, it is important to recognize that these probabilities apply to both transmittances and area fractions. Binomial probability distributions are not particularly easy to work with, so we make another assumption to simplify the math; namely, that the area of a single leaf a is much smaller than the ground area, A. This certainly is reasonable for many realistic examples. If the mathematical limit of $(1 - a/A)^N$ is taken as $a/A \rightarrow 0$, then $(1 - a/A)^N \simeq \exp(-Na/A)$, where Na/A is the leaf area index. Essentially this amounts to replacing the binomial distribution with the Poisson distribution. Interestingly, this result for horizontal leaves is independent of the incident angle of the light, because the shadow of a horizontal leaf cast onto a horizontal plane is the area of the leaf no matter what direction the light comes from.

A quick look at vegetation reveals that leaves are almost never oriented entirely horizontally; they have a distribution of orientations. However, the horizontal shadows of black leaves can always be treated as "horizontal leaves" no matter what the leaf orientation. Although these shadows have a range of sizes, they all are random relative to each other and small compared to the area, A. This is all that is required for exponential extinction to hold. Therefore, no matter what the orientation distribution of leaves, the fraction of the leaf HSAI that is projected onto the horizontal plane from a particular zenith angle ψ can be calculated: We refer to this fraction as the extinction coefficient, $K_b(\psi)$. One can also think of $K_b(\psi)$ as the mean beam flux density on an average illuminated leaf in the canopy divided by the beam flux density on the horizontal plane above the canopy. Clearly, for a canopy of perfectly horizontal leaves, $K_b(\psi) = 1$. The azimuth angle of the incident radiation may also be important in estimating $K_b(\psi)$, but we deal only with canopies that have leaves symmetrically distributed about the azimuth (compass directions), which is a good assumption for almost all canopies. The extinction coefficient is therefore assumed independent of solar azimuth. Later we calculate $K_b(\psi)$ for various leaf-angle distributions and remove the assumption of "black" leaves.

If flat leaves in a canopy of leaf area index L_t are randomly distributed in space, then the fraction $\tau_b(\psi)$ of incident beam radiation from zenith angle ψ that penetrates the canopy is

$$\tau_b(\psi) = \exp(-K_b(\psi)L_t) \qquad (15.1)$$

where $K_b(\psi)$ is the canopy extinction coefficient just described. When leaves are clumped (not randomly distributed), canopy transmission can often still be approximated by an exponential function of L_t, but L_t is multiplied by a clumping factor Ω to account for the fact that leaves are less efficient in covering the ground than when they are randomly distributed. Row crops with leaves clumped in the rows may intercept only 70 to 80 percent of the radiation they would if their leaves were randomly distributed in space.

The fraction of incident beam radiation intercepted by the canopy is $1 - \tau_b(\psi)$. This fraction is available for scattering, transpiration, and for photosynthesis. The fraction of beam radiation that is not intercepted by the canopy ($\tau_b(\psi)$) reaches the soil surface and is available for evaporating water from the soil. A simple way to partition potential evapotranspiration (PET) between potential transpiration and potential soil evaporation uses τ. Potential transpiration is $1 - \tau$ times PET, and potential evaporation is τ times PET. A canopy that covers the ground reasonably well has a leaf area index of perhaps three. If $K(\psi) = 0.6$, then, from Eq. (15.1), $\tau(\psi) = \exp(-0.6 \times 3) = 0.17$; so 17 percent of the radiation is intercepted by the soil surface and 83 percent is intercepted by the canopy. If both canopy and soil surface were wet, so that evapotranspiration were at the potential rate, then 83 percent of the evapotranspiration would come from the canopy and 17 percent from the soil.

15.2 Detailed Models of Light Interception by Canopies

Our purpose here is to find equations that allow us to account for the major variations in PAR and near-infrared (NIR) fluxes on leaves in a canopy. The most obvious variations result from shading of some leaves by others. We therefore consider two classes of leaves, those that are shaded, and those that are sunlit. Average PAR or NIR flux densities for each of these classes can be derived. More detailed models subdivide each of these classes to account for the leaf angle distribution and position in the canopy of leaves, but we do not consider those now. Goudriaan (1988) shows how to derive a model with more radiation classes.

The calculation of an extinction coefficient requires calculating the area of an average projection from some direction ψ onto the horizontal, and this is not an easy thing, except by some geometrical reasoning. If all of the leaves in a canopy were vertical, but with random azimuthal orientations, then the distribution function for leaf area in the canopy would be the same as the distribution function for area on the vertical surface of a vertical cylinder. The ratio of the area projected onto the horizontal from the direction ψ to the hemi-surface area of a cylinder (length L_c and diameter D) is the extinction coefficient and it is given by

$$K_{bc}(\psi) = \frac{L_c D \tan \psi}{\frac{\pi}{2} DL_c} = \frac{2 \tan \psi}{\pi} \qquad (15.2)$$

where ψ is the zenith angle of the sun. Similarly, a crop might have leaves with leaf inclination angles similar to the distribution of angles on the surface of a sphere. Taking the ratio of the area of the projection of a sphere (radius r) onto a horizontal surface to the hemi-surface area of the sphere gives

$$K_{bs}(\psi) = \frac{\frac{\pi r^2}{\cos \psi}}{2\pi r^2} = \frac{1}{2 \cos \psi}. \qquad (15.3)$$

A canopy with a spherical leaf angle distribution does not need to look like a ball. Imagine cutting the surface of a sphere into many little pieces, then moving these pieces about the volume occupied by the canopy while maintaining the zenith and azimuth orientations of each piece. The resulting canopy would have a spherical angle distribution. There would be more vertical area than horizontal area, because more of the surface area of a sphere is vertical than horizontal, but leaves of all inclinations would be present in the canopy. A spherical angle distribution is a good approximation to real plant canopies.

An extinction coefficient for a conical leaf distribution could also be derived, but the most useful distribution is ellipsoidal. The ellipsoidal distribution generalizes the spherical, but allows the sphere to be flattened or elongated. The ratio of projected area to hemi-surface area for an ellipsoid is (Campbell, 1986):

$$K_{be}(\psi) = \frac{\sqrt{x^2 + \tan^2 \psi}}{x + 1.774(x + 1.182)^{-0.733}}. \tag{15.4}$$

Here, the parameter x is the ratio of average projected areas of canopy elements on horizontal and vertical surfaces. For a spherical leaf angle distribution, $x = 1$; for a vertical distribution, $x = 0$; and for a horizontal leaf canopy, x approaches infinity. Equation (15.4) therefore gives all of the simple K_b's and all of the ones in between. Figure 15.1 shows leaf angle density for three different values of x. The equation for these distributions is given by Campbell (1990). As mentioned, the spherical distribution has more vertical than horizontal area, but spreads the area fairly uniformly among almost all angles. As x increases the peak shifts toward horizontal angles and as x decreases the peak shifts toward vertical angles. If we were to plot the horizontal and vertical distributions on Fig. 15.1, they would be infinitesimally narrow, and infinitely tall spikes (called Dirac delta functions) at 0 and 90°.

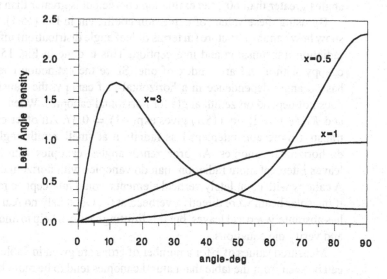

FIGURE 15.1. Inclination angle density for three canopies. The larger x is, the more horizontal the leaves are.

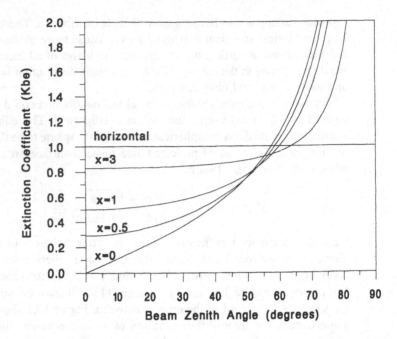

FIGURE 15.2. The extinction coefficient $K_{be}(\psi)$ as a function of zenith angle for x values representing various leaf angle distributions.

Figure 15.2 shows extinction coefficients as a function of beam zenith angle for a range of x values. Note that extinction in horizontal canopies has no zenith angle dependence, but for all other canopies, zenith angles below about 60° have extinction coefficients below unity, while at zenith angles greater than 60°, the extinction coefficient is greater than unity.

By using these values of extinction coefficient in Eq. (15.1), we can show how canopy structure (in terms of leaf angle distribution) influences radiation transmission and interception. This is done in Fig. 15.3 for a canopy with a leaf area index of one. Since the extinction coefficient has no angle dependence in a horizontal-leaf canopy, the transmission does not depend on zenith angle for horizontal canopies. When $L_t = 1$, and $K_{be}(\psi) = 1$, Eq. (15.1) gives $\exp(-1) = 0.37$. All other canopies transmit more and intercept less radiation at small zenith angles than do horizontal canopies. At large zenith angles, canopies with inclined leaves intercept more radiation than do canopies with horizontal leaves. A canopy with completely vertical elements would intercept no radiation if the solar beam were directly overhead at 0°. Obviously no real canopy has absolutely vertical leaves, but this limiting case can help to understand and verify the equations.

Measured values of x for a number of crops are given in Table 15.1. It can be seen from the table that natural canopies tend to be more horizontal than vertical and that the spherical distribution ($x = 1$) approximates many of the canopies. If no information is available about the angle distri-

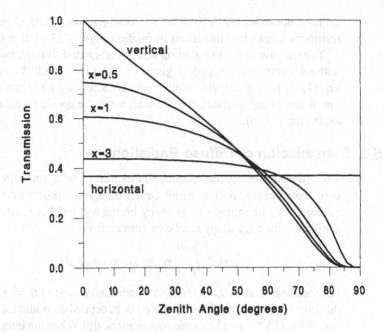

FIGURE 15.3. Fraction of the incident solar beam reaching the ground below a canopy with LAI = 1, for different leaf angle distributions.

bution of leaves in the canopy, it is often assumed to be spherical. Values of x are not particularly intuitive for understanding leaf orientation. Mean leaf inclination angle is more easily understood. The mean leaf inclination angle can be approximated as $\cos^{-1}(K_{be}(0))$ using Eq. (15.4). The mean leaf inclination angles are therefore about $73°$, $60°$, and $34°$ for x values of 0.5, 1.0, and 3.0. This approximation of mean leaf inclination

TABLE 15.1. Values of the leaf angle distribution parameter x for various crop canopies (from Campbell and van Evert, 1994)

Crop	x	Crop	x
Ryegrass	0.67–2.47	Cucumber	2.17
Maize	0.76–2.52	Tobacco	1.29–2.22
Rye	0.8–1.27	Potato	1.70–2.47
Wheat	0.96	Horse Bean	1.81–2.17
Barley	1.20	Sunflower	1.81–4.1
Timothy	1.13	White clover	2.47–3.26
Sorghum	1.43	Strawberry	3.03
Lucerne	1.54	Soybean	0.81
Hybrid swede	1.29–1.81	Maize	1.37
Sugar beet	1.46–1.88	J. artichoke	2.16
Rape	1.92–2.13		

angle is not exact but it is close. For example, the spherical leaf angle distribution has a true mean leaf inclination angle of 57°, rather than 60°.

The fraction of beam radiation that is transmitted through the canopy without interception $\tau_b(\psi)$ is given by Eq. (15.1) with $K_{be}(\psi)$ from Eq. (15.4) (or one of the simpler equations for $K_b(\psi)$ if the distribution is horizontal, vertical, or spherical) using the appropriate sun zenith angle (Eq. (11.1)).

15.3 Transmission of Diffuse Radiation

The diffuse radiation comes from all directions, and is attenuated differently from beam radiation, which comes from just one direction. Diffuse radiation can be thought of as many beams and a diffuse transmission coefficient for the canopy can be calculated from

$$\tau_d = 2 \int_0^{\pi/2} \tau_b(\psi) \sin \psi \cos \psi \, d\psi. \tag{15.5}$$

For horizontal leaves, $\tau_b(\psi)$ is not dependent on ψ, and so $\tau_b = \tau_d$, but for the other leaf angle distributions $\tau_b(\psi)$ does depend on ψ and the integration in Eq. (15.5) must be carried out numerically. When the integration is done numerically, it is found that τ_d does not decrease exponentially with L, as it does for beam radiation (except for horizontal leaves). In order to obtain a useful approximation for models, an exponential equation can be fit to the values obtained and allow K_d, the extinction coefficient for black leaves in diffuse radiation, to vary with leaf area index. Figure 15.4 shows the result based on a numerical integration of Eq. (15.5), assuming

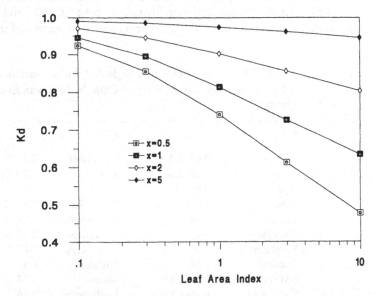

FIGURE 15.4. Apparent extinction coefficient for diffuse radiation in canopies differing in leaf angle distribution.

a uniform overcast sky (no zenith angle dependence of sky radiance). For horizontal leaves, $K_d = 1$, but for a spherical canopy, with $L_t = 3$, K_d is around 0.7. This is an important point which we return to later.

Note that diffuse radiation, unlike beam radiation from the sun, is distributed relatively uniformly over all leaves with various orientations for a particular layer in the canopy. Thus the diffuse flux density incident on a leaf at some depth L in the canopy is the same as the diffuse flux density estimated on the horizontal at the same depth using Eq. (15.5) and the diffuse flux density above the canopy.

15.4 Light Scattering in Canopies

The leaves in plant canopies are not black, of course, and do transmit and reflect radiation. Goudriaan (1977) has shown that the transmission and reflection of radiation when the leaves are not assumed black can still be approximated using an exponential model (Eq. (15.1)), but with a modification to K. If the absorptivity of leaves for radiation is α, then the total beam radiation (direct and down scattered) transmitted through the canopy to depth L is

$$\tau_{bt}(\psi) = \exp(-\sqrt{\alpha}K_{be}(\psi)L). \tag{15.6}$$

It can be seen that when $\alpha = 1$ (black leaves) this equation is the same as Eq. (15.1) and when α is small radiation will be attenuated minimally. The transmission of light through the leaves therefore gives an additional amount of radiation under the canopy. Equation (15.6) is an approximation, and Goudriaan (1977) has shown (his Table 5, p. 27) that Eq. (15.6) works well for a range of sun zenith angles, canopy architectures, and leaf absorptivity values. For a canopy with a spherical leaf angle distribution, Eq. (15.6) works well for sun zenith angles less than 65°. The transmission of diffuse radiation by the canopy is predicted by a similar equation, but with K_d as the extinction coefficient. Typical values for α are $\alpha_p = 0.8$ for PAR and $\alpha_n = 0.2$ for NIR radiation. For total solar radiation, absorptivity is the mean of the values for PAR and NIR, so $\alpha_s = 0.5$.

15.5 Reflection of Light by Plant Canopies

For a canopy of randomly located, horizontally oriented leaves with a LAI so large that the soil has negligible effect on radiation reflected from the canopy, the canopy hemispherical reflection coefficient, ρ_{cpy}^H, is given by

$$\rho_{cpy}^H = \frac{1 - \sqrt{\alpha}}{1 + \sqrt{\alpha}} \tag{15.7}$$

where α is the leaf absorptivity. This means that for a dense canopy of horizontal leaves, in the PAR ($\alpha = 0.8$), $\rho_{p,cpy}^H = 0.056$; in the NIR ($\alpha = 0.2$), $\rho_{N,cpy}^H = 0.38$; and in the solar ($\alpha = 0.5$), $\rho_{S,cpy}^H = 0.17$. This canopy reflection coefficient for solar radiation actually is not a reliable estimate. Equation (15.6) accommodates multiple scattering in the canopy and is only appropriate where reflectivity and transmissivity are constant with wavelength. If the reflection coefficient is averaged over a wavelength band where spectral reflectivity (and transmissivity) varies considerably with wavelength, (as it does for leaves in the visible and near-infrared portions of the solar spectrum) then Eq. (15.7) is unreliable. This can best be understood with a simple example; shown as Example 15.1.

Example 15.1. Estimate the transmission of solar radiation through two filters, stacked on top of each other, using the following two methods.
1. Assume an average transmission for the solar wavelength band (τ_S),
2. Use visible (VIS) and near-infrared (NIR) transmissions separately.

Assume 1/2 of the solar radiation is NIR and 1/2 is VIS, the visible transmittance (τ_V) is 0.0, and near-infrared transmittance (τ_N) is 0.9.

Solution.
1. Using τ_S:

$$\tau_S = 0.5(0) + 0.5(0.9) = 0.45$$

$$\tau(2\text{ filters}) = \tau_S\tau_S = 0.45 \times 0.45 = 0.20.$$

2. Using τ_V and τ_N with $\tau_V = 0$ and $\tau_N = 0.9$:
   visible

$$\tau(2\text{ filters}) = 0.0 \times 0.0 = 0.0$$

near-infrared

$$\tau(2\text{ filters}) = 0.9 \times 0.9 = 0.81$$

solar

$$\tau(2\text{ filters}) = 0.5(0.) + 0.5(0.81) = 0.40.$$

Therefore averaging multiple transmissions or reflections, as happens in plant canopies, over wavelength bands with different spectral properties causes errors; in this example a factor of two.

From Example 15.1, the visible and near-infrared wavelength bands should be treated separately because their spectral properties are so different. Thus a better estimate of the solar albedo is given by $\rho_{S,\text{cpy}}^{H} = 0.5(0.056) + 0.5(0.38) = 0.22$, a value 29 percent larger than is obtained by substituting the average solar absorptivity into Eq. (15.7). This is one of the reasons that solar radiation must be divided into visible and near-infrared wavelength bands in environmental biophysics. Fortunately about one-half of the irradiance is in each band so approximate partitioning is simple.

If the leaves are not horizontal, Goudriaan (1988) suggests that the beam reflection coefficient for a deep canopy can be approximated from

$$\rho_{b,\text{cpy}}^*(\psi) = \frac{2K_{be}(\psi)}{K_{be}(\psi) + 1}\, \rho_{\text{cpy}}^H. \qquad (15.8)$$

The reflection coefficient for diffuse radiation can be approximated by substituting K_d for $K_{be}(\psi)$ in Eq. (15.8).

If the canopy is not dense, then the effect of the soil may be significant and the canopy reflection coefficient for beam irradiance becomes (Monteith and Unsworth, 1990)

$$\rho_{b,\text{cpy}}(\psi) = \frac{\rho_{b,\text{cpy}}^* + \left[\frac{\rho_{b,\text{cpy}}^* - \rho_s}{\rho_{b,\text{cpy}}^* \rho_s - 1}\right]\exp(-2\sqrt{\alpha}K_{be}(\psi)L_t)}{1 + \rho_{b,\text{cpy}}^*\left[\frac{\rho_{b,\text{cpy}}^* - \rho_s}{\rho_{b,\text{cpy}}^* \rho_s - 1}\right]\exp(-2\sqrt{\alpha}K_{be}(\psi)L_t)}. \qquad (15.9)$$

Neglecting second order terms like $(\rho_{b,\text{cpy}}^*(\psi))^2$ and $\rho_{b,\text{cpy}}^*(\psi)\rho_s$ results in the following simplified equation:

$$\rho_{b,\text{cpy}} \simeq \rho_{b,\text{cpy}}^* - (\rho_{b,\text{cpy}}^*(\psi) - \rho_s)\exp(-2\sqrt{\alpha}K_{be}(\psi)L_t). \qquad (15.10)$$

Equation (15.10) is a good approximation to Eq. (15.9) in the PAR, but in the NIR, relative discrepancies can approach five percent. The diffuse forms of Eqs. (15.9) and (15.10) have $K_{be}(\psi)$ replaced by K_d and are represented by $\rho_{d,\text{cpy}}$. ρ_s is soil reflectance.

15.6 Transmission of Radiation by Sparse Canopies—Soil Reflectance Effects

For a canopy with a high LAI, the transmission of beam radiation (including its scattered component) as a function of depth L in the canopy is given by Eq. (15.6). If the canopy is not dense, and the LAI is low, then radiation can be reflected from the soil and re-reflected from the leaves to enhance the downwelling radiation stream. Monteith and Unsworth (1990), give the following equation for determining the flux density of radiation under the canopy:

$$\tau_{bt}(\psi) = \frac{\left[(\rho_{b,\text{cpy}}^*(\psi))^2 - 1\right]\exp(-\sqrt{\alpha}K_{be}(\psi)L_t)}{(\rho_{b,\text{cpy}}^*(\psi)\rho_s - 1) + \rho_{b,\text{cpy}}^*(\psi)(\rho_{b,\text{cpy}}^*(\psi) - \rho_s)\exp(-2\sqrt{\alpha}K_{be}(\psi)L_t)}. \qquad (15.11)$$

If the second order terms are again neglected, then Eq. (15.11) simplifies to Eq. (15.6), and this amounts to assuming that the ratio of upwelling to downwelling radiation below L_t for a deep canopy is equivalent to the soil reflectance for a finite canopy. In the PAR wavelength band, Eq. (15.6) may be a reasonable approximation to Eq. (15.11), depending on ρ_s, but in the NIR, relative descrepancies of ten percent or more can occur. The beam radiation absorbed by the canopy can be approximated with

$$\alpha_{b,\text{cpy}}(\psi) = 1 - \rho_{b,\text{cpy}}(\psi) - \tau_{b,\text{cpy}}(\psi)(1 - \rho_s). \qquad (15.12)$$

while the beam radiation intercepted by the canopy is

$$f_b = 1 - \tau_b(\psi) \tag{15.13}$$

Clearly the absorptivity of the canopy depends on wavelength but the interception does not depend on wavelength.

15.7 Daily Integration

Equation (14.13) requires estimates of the fraction of radiation intercepted by the canopy, averaged over whole days. Fuchs et al. (1976) suggested that the interception of beam and diffuse radiation, averaged over whole days, can be approximated by the intercepted function for diffuse radiation because the sun traverses the whole sky over the period of the day. Tests with detailed models have shown this to be correct. Therefore the average transmission of canopies can be modelled over whole days using Eq. (15.6), with $K_{be}(\psi)$ replaced by K_d (from Fig. 15.4).

Based on these observations, the daily fractional interception can be computed from

$$f = 1 - \exp(-K_d L_t). \tag{15.14}$$

Absorption of PAR is about equal to interception, while absorption of total solar radiation is about 80 percent of interception (Campbell and van Evert, 1994).

15.8 Calculating the Flux Density of Radiation on Leaves in a Canopy

The equations we have just derived can be used to compute the flux density of radiation on leaves within the canopy. Knowing the flux density on leaves is important for the purpose of computing photosynthesis and for calculating the radiation viewed by a remote sensor.

Let Q_{ob} be the flux density of beam radiation on a horizontal surface at the top of the canopy and Q_{od} be the flux density of diffuse radiation on the horizontal above the canopy. At a depth L in the canopy, three different flux densities can be calculated: the total beam, $Q_{bt}(\psi)$ (unintercepted beam plus down scattered beam); beam, $Q_b(\psi)$ (unintercepted beam) and the diffuse flux, Q_d. These are given by

$$Q_{bt}(\psi) = \tau_{bt}(\psi)Q_{ob} \tag{15.15}$$

$$Q_b(\psi) = \tau_b(\psi)Q_{ob} \tag{15.16}$$

$$Q_d = \tau_{dt}Q_{od}. \tag{15.17}$$

Here, $\tau_{bt}(\psi)$ and τ_{dt} are given by Eq. (15.6) with the appropriate K for beam or diffuse radiation, and $\tau_b(\psi) = \exp(-K_{be}(\psi)L)$.

At depth L in the canopy some leaves are sunlit and some leaves are in the shade. The flux density on a horizontal surface at the position of a sunlit leaf is $Q_{bt}(\psi) + Q_d$. The flux density on the leaves themselves will vary depending on their orientation, but the mean flux density on the sunlit leaves can be shown to be

$$Q_{sl}(\psi) = K_{be}(\psi)Q_{ob} + Q_d + Q_{sc} \qquad (15.18)$$

where Q_{sc} is the flux density of down-scattered radiation from the solar beam. The flux density on shaded leaves is the diffuse flux plus the down-scattered flux from the solar beam:

$$Q_{sh} = Q_d + Q_{sc}. \qquad (15.19)$$

The down-scattered radiation is the difference between $Q_{bt}(\psi)$ and $Q_b(\psi)$:

$$Q_{sc} = Q_{bt}(\psi) - Q_b(\psi). \qquad (15.20)$$

The next problem is to know what fraction of the leaf area at depth L is sunlit. The probability of finding a sunlit leaf area index in thickness δL at depth L in the canopy is the product of the probability that a ray will penetrate to depth L and the probability that it will be intercepted in the layer δL, divided by $K_{be}(\psi)$ (the ratio of projections of leaf area on a horizontal surface to actual leaf area). If L^* is used to represent the sunlit leaf area index, then

$$\delta L^* = \frac{\exp(-K_{be}(\psi)L)\left[1 - \exp(-K_{be}(\psi)\delta L)\right]}{K_{be}(\psi)}. \qquad (15.21)$$

In the limit as δL becomes small, $\delta L^* = \delta L \exp(-K_{be}(\psi)L)$. The fraction $f_{sl}(\psi)$ of sunlit leaves at depth L is $\delta L^*/\delta L$, so

$$f_{sl}(\psi) = \exp(-K_{be}(\psi)L) = \tau_b(\psi). \qquad (15.22)$$

The fraction of shaded leaves is $f_{sh}(\psi) = 1 - f_{sl}(\psi)$. If the LAI of the entire canopy is L_t, then the sunlit LAI of the whole canopy L_t^* is

$$L_t^* = \frac{1 - \exp\left[-K_{be}(\psi)L_t\right]}{K_{be}(\psi)} \qquad (15.23)$$

and the shaded LAI is $L_t - L_t^*$.

15.9 Calculating Canopy Assimilation from Leaf Assimilation

Several methods are available for calculating canopy photosynthetic rate from leaf photosynthetic rate based on the distribution of light over leaves, including methods that consider additional factors such as wind and humidity. Norman (1992) compared various simple methods for estimating canopy assimilation from leaf assimilation. The most robust method seems to divide the canopy into sunlit and shaded leaf classes, calculate the assimilation rate for representative members of each class, and sum the two contributions according to the fraction of leaf area in each class. One reason this method works so well is that it accommodates the nonlinear response of leaf assimilation to light. Light assimilation responses of leaves (see Fig. 14.6) can vary with depth in the canopy and this variation can be accommodated by partitioning the canopy into several layers and estimating the sunlit and shaded leaf fractions in each layer. Usually this is not necessary and a single, representative light assimilation response curve can be used for the entire canopy. Obviously most of the sunlit leaves are near the top of the canopy and most of the shaded leaves are near the bottom; therefore, one minor adjustment might be to use slightly different light assimilation response curves for sunlit and shaded leaves. In our example we use a single light assimilation response relation for all the leaves in the canopy.

Example 15.2. Estimate the canopy photosynthetic rate at 10°C (light assimilation curve in Fig. 14.6) for a canopy with a spherical leaf angle distribution and hemi-surface area index of 3.0, incident PAR above the canopy on the horizontal of $Q_{ob} = 2000$ μmol photons m^{-2} s^{-1} with 80 percent as beam and 20 percent as diffuse radiation, sun zenith angle $\psi = 40°$, and leaf absorptivity $\alpha = 0.8$.

Solution. The canopy net assimilation rate $A_{n,cpy}$ is the sum of contributions of sunlit and shaded leaves. These two contributions are added separately because sunlit leaves will be light saturated while shaded leaves will be in the linear portion of the light assimilation relation; thus canopy assimilation is not proportional to average light levels:

$$A_{n,cpy} = A_{n,leaf}^{sun} L_t^* + A_{n,leaf}^{shade}(L_t - L_t^*) \qquad (15.24)$$

where $A_{n,leaf}$ is the μmol CO_2 m^{-2} (leaf hemi-surface area) s^{-1}, and $A_{n,cpy}$ is the μmol CO_2 m^{-2} (ground area) s^{-1}, and, of course, L_t and L_t^* are (leaf hemi-surface area)(ground area)$^{-1}$.

First the average PAR incident on shaded leaves needs to be estimated. At the top of the canopy shaded leaves receive the diffuse radiation from the sky, 400 μmol photons m^{-2} (ground area) s^{-1}. At the bottom of the canopy $Q_{sh} = Q_d + Q_{sc}$. From Fig. 15.4, $K_d = 0.72$, so:

$$Q_d = \tau_{dt} Q_{od} = Q_{od} \exp(-\sqrt{\alpha} K_d L_t)$$

$$= 400 \exp(-\sqrt{0.8} x 0.72 x 3.0)$$

$$= 58 \ \mu\text{mol photons m}^{-2} \text{ (ground area)s}^{-1}.$$

The diffuse PAR on a horizontal plane is 400 at the top and 58 μmol photons m^{-2} (ground area) s^{-1} at the bottom. For diffuse radiation, the flux density on the horizontal is assumed the same as the flux density on a leaf if the leaf area is expressed on a hemi-surface area basis (leaf HSA). Thus top shaded leaves have a diffuse illumination of 400 and bottom leaves receive 58 μmol photons m^{-2} (leaf hemi-surface area) s^{-1}. These two values could be averaged to obtain 229 μmol photons m^{-2} (leaf hemi-surface area) s^{-1}, but it is known that the attenuation is exponential and not linear, so a more appropriate average is an exponentially-weighted average:

$$\overline{Q_d} = \frac{\int_0^{L_t} Q_d dL}{\int_0^{L_t} dL} = \frac{Q_{od}\left[1 - \exp(-\sqrt{\alpha}K_d L_t)\right]}{\sqrt{\alpha}K_d(L_t - 0)}$$

$$= 177 \frac{\mu\text{mol}}{\text{m}^2(\text{leaf hemi-surface area}) \text{ s}}.$$

The scattered beam radiation is zero at the top of the canopy and is given by $Q_{sc} = Q_{bt}(\psi) - Q_b(\psi)$ at the bottom. If the beam extinction coefficient is

$$K_{be}(40°) = \frac{\sqrt{(1^2 + 0.839^2)}}{1 + 1.774(1 + 1.182)^{-.733}} = 0.652,$$

then

$$Q_{bt} = 1600 \exp(-\sqrt{0.8} \times 0.652 \times 3.0) = 1600 \times (0.174)$$

$$= 278 \; \mu\text{mol photons m}^{-2} \text{ (ground area) s}^{-1}$$

$$Q_b = 1600 \exp(-0.652 x 3.0) = 1600 x 0.141$$

$$= 226 \; \mu\text{mol photons m}^{-2} \text{ (ground area) s}^{-1}$$

$$Q_{sc} = 278 - 226 = 52 \; \mu\text{mol photons m}^{-2} \text{ (ground area)s}^{-1}.$$

Therefore the average scattered illumination on leaves is $(52 + 0)/2 = 26 \; \mu\text{mol photons m}^{-2}$ (leaf hemi-surface area) s^{-1}. The PAR flux density absorbed by shaded leaves is

$$\overline{Q_{sh}} = 0.8(177 + 26) = 0.8 \times 203$$

$$= 162 \frac{\mu\text{mol photons}}{\text{m}^2(\text{leaf hemi-surface area}) \text{ s}}$$

where the overbar denotes an average over the depth of the canopy. The PAR flux density absorbed by sunlit leaves is given by

$$\overline{Q_{sl}} = \alpha(K_{be}(\psi)Q_{ob} + \overline{Q_{sh}})$$

$$\overline{Q_{sl}} = 0.8(0.652 \times 1600 + 203) = 0.8 \times 1246$$

$$= 997 \frac{\mu\text{mol photons}}{\text{m}^2(\text{leaf hemi-surface area}) \text{ s}}$$

The sunlit LAI (L_t^*) and shaded LAI ($L_t - L_t^*$) are given by

$$L_t^* = \frac{1 - \exp(-K_{be}(40°)L_t)}{K_{be}(40°)} = \frac{1 - 0.141}{0.652}$$

$$= 1.32 \frac{\text{m}^2(\text{leaf hemi-surface area})}{\text{m}^2(\text{ground area})}$$

$$L_t - L_t^* = 3.0 - 1.32 = 1.68 \frac{\text{m}^2(\text{leaf hemi-surface area})}{\text{m}^2(\text{ground area})}.$$

The leaf assimilation rates can be obtained from Fig. 14.6 using $\overline{Q_{sh}}$ for shaded leaves and $\overline{Q_{sl}}$ for sunlit leaves. The leaf assimilation rates in Fig.

14.6 are in units of μmol CO_2 m^{-2}(leaf surface area) s^{-1}, and the units needed are μmol CO_2 m^{-2}(leaf hemi-surface area) s^{-1}. Therefore

$$A_{n,\text{leaf}}^{\text{sun}} = 11 \frac{\mu\text{mol } CO_2}{\text{m}^2(\text{leaf surface area}) \text{ s}}$$

$$\times \frac{2\text{m}^2(\text{leaf surface area})}{\text{m}^2(\text{leaf hemi-surface area})}$$

$$A_{n,\text{leaf}}^{\text{sun}} = 22 \frac{\mu\text{mol } CO_2}{\text{m}^2(\text{leaf hemi-surface area}) \text{ s}}.$$

Similarily, at $162/2 = 81$ μmol photons m^{-2} (leaf surface area) s^{-1}

$$A_{n,\text{leaf}}^{\text{shade}} = 6 \frac{\mu\text{mol } CO_2}{\text{m}^2(\text{leaf hemi-surface area}) \text{ s}}.$$

Therefore the canopy assimilation is given by

$$A_n = 22 \frac{\mu\text{mol } CO_2}{\text{m}^2(\text{leaf hemi-surface area}) \text{ s}}$$

$$\times 1.32 \frac{\text{m}^2(\text{leaf hemi-surface area})}{\text{m}^2(\text{ground area})}$$

$$+ 6 \frac{\mu\text{mol } CO_2}{\text{m}^2(\text{leaf hemi-surface area}) \text{ s}}$$

$$\times 1.68 \frac{\text{m}^2(\text{leaf hemi-surface area})}{\text{m}^2(\text{ground area})}$$

$$= 29.0 + 10.1 = 39.1 \frac{\mu\text{mol } CO_2}{\text{m}^2(\text{ground area}) \text{ s}}.$$

The approach used in Example 15.2 to scale leaf assimilation to canopy assimilation accommodates the nonlinearity in the 10°C light assimilation curve in Fig. 14.6. If we had ignored the fact that the 10°C light assimilation curve is not a straight line and used an average absorbed PAR for the entire canopy to scale up the leaf assimilation rate, how large would the error be? The average absorbed PAR for the canopy $\overline{Q}$ is the mean of sunlit and shaded absorbed PAR weighted by the leaf area of each:

$$\overline{Q} = \frac{\overline{Q_{sl}}L_t^* + \overline{Q_{sh}}(L_t - L_t^*)}{L_t} = \frac{997 \times 1.32 + 162 \times 1.68}{3.0}$$

$$= 529 \frac{\mu \text{ mol photons}}{\text{m}^2 \text{ (leaf hemi-surface area) } s}.$$

From Fig. 14.6, the leaf assimilation rate corresponding to the average absorbed PAR is 20 μmol CO_2 m^{-2}(leaf hemi-surface area) s^{-1}, so the canopy assimilation rate is given by

$$
A_{n,\text{cpy}} \cong 20 \, \frac{\mu\text{mol CO}_2}{\text{m}^2(\text{leaf hemi-surface area}) \, s}
$$

$$
\times \, 3 \, \frac{\text{m}^2(\text{leaf hemi-surface area})}{\text{m}^2(\text{ground area})}
$$

$$
\cong 60 \, \frac{\mu\text{mol CO}_2}{\text{m}^2(\text{ground area}) \, s}.
$$

This is 54 percent larger than the sunlit/shaded method in Example 15.2. This value of 60 μmol CO_2 m^{-2}(ground area) s^{-1} is approximately the canopy photosynthetic rate that would occur if the canopy were illuminated with entirely diffuse irradiance at 2000 μmol photons m^{-2}(ground area) s^{-1}. Thus a diffuse irradiance of about 1300 μmol photons m^{-2}(ground area) s^{-1} would result in about the same canopy photosynthetic rate (39 μmol CO_2 m^{-2} (ground area) s^{-1}) as 2000 μmol photons m^{-2}(ground area) s^{-1} with 80 percent beam and 20 percent diffuse. This means that diffuse irradiance is more efficient for photosynthesis than beam irradiance.

Light assimilation responses are not always as nonlinear as the 10°C curve in Fig. 14.6; for example, the 30°C curve in Fig. 14.6. Comparing the canopy assimilation prediction from the sunlit/shaded method with the average-APAR method results in 39 μmol CO_2 m^{-2} (ground area) s^{-1} for both methods; this occurs because of the linearity of the 30°C curve. Considering the greater leaf assimilation rate at 30°C from Fig. 14.6, it may be surprising to find the canopy assimilation rates for 10 and 30°C are nearly equal. This occurs because leaves at 30°C have higher photosynthetic rates on sunlit leaves and lower rates on shaded leaves, because of the larger dark respiration. Essentially the higher maximum leaf photosynthetic rate comes at a higher dark respiration cost. Furthermore, the canopy architecture limits the fraction of leaves that can be sunlit.

Leaf stomatal conductance can be scaled to a canopy conductance by using the same method as outlined above for photosynthetic rates if stomatal conductances for sunlit and shaded leaves are known. Using stomatal conductances appropriate for the leaf assimilation rates plotted in Fig. 14.6 under humid atmospheric conditions, the canopy conductance for Example 15.2 can be estimated from an equation like Eq. (15.24) to be 0.5(1.32) + 0.2(1.68) = 1.0 mol water m^{-2}(ground area) s^{-1}. Because sunlit LAI approaches a maximum as LAI increases (L_t^* has a maximum of about 1.5 for high LAI canopies with $\psi = 40°$) and the mean shaded stomatal conductance decreases as LAI increases, this sunlit/shaded approach clearly shows why canopy conductances tend to reach maximum values that might be expected to be related to sunlit leaf area index.

15.10 Remote Sensing of Canopy Cover and IPAR

Remote sensing is a name associated with inferring characteristics of surfaces from measurements of radiance. In environmental biophysics, remote sensing usually refers to the interpretation of radiometric measurements made above soil–vegetation systems from towers, aircraft, or satellites. A more general term is indirect measurement, which refers to any measurement made without directly contacting an object. Technically, our eyes indirectly sense the environment around us so an absurd interpretation might infer that all information obtained with our eyes (e.g., reading a ruler) could be considered remote sensing; however, this is not what we mean. In environmental biophysics, some examples of remote sensing include the following.

1. Infrared thermometer measurements of soil surface temperature.
2. Measuring soil or canopy roughness using the backscattered radiation from a laser (these systems are called LIDAR).
3. Estimating the water content of the top 5 cm layer of soil using passive microwave measurements of surface temperature and emissivity.
4. Estimating total forest-canopy water content to infer vegetation biomass using RADAR.
5. Inferring canopy cover, leaf area index, or intercepted photosynthetically active radiation (IPAR) from measurements of visible (VIS) and near-infrared (NIR) reflected radiance.

Another indirect measurement that is common in environmental biophysics, but not generally referred to as remote sensing, is the indirect measurement of canopy architecture. This is discussed briefly in a later section of this chapter.

Some of the fundamental characteristics of remote sensing data can be understood using knowledge of canopy architecture by considering the relation between canopy cover, IPAR, and reflected VIS and NIR radiation. In previous sections we discussed the penetration of radiation through canopies, the reflection of radiation from canopies and the distribution of radiation over the surface of leaves. Although all this is relevant to remote sensing, a second consideration also is required; that is, the portion and characteristics of the canopy and soil that occupy the field-of-view (FOV) of the sensor. As mentioned in Ch. 10, bidirectional reflectance factors (BRF) involve two directions; the direction of the source (usually the sun) and the direction of the receiver (a sensor). To simplify the analysis that follows, we do not consider finite solid angles of view, but only consider particular directions as though the radiation were composed of parallel rays all from that direction. Essentially this amounts to using data from a narrow FOV sensor that is calibrated to read out the flux density eminating from the target surface by making the output proportional to the radiance times the FOV of the sensor.

For our purposes, the bidirectional reflectance factor (BRF) for a surface can be defined as follows:

$$\text{BRF} = \frac{\text{flux density leaving the horizontal surface viewed by a sensor}}{\text{flux density incident on the horizontal surface}}.$$

The flux density incident on the surface usually is measured by pointing the sensor at a reference surface (exposed to the same illumination conditions as the target surface) that is as close to a perfectly-reflecting, Lambertian surface as possible. Clearly the BRF may be different for various wavelength bands such as the visible (BRF_V) and near-infrared (BRF_N), and the view of the sensor may be occupied by sunlit leaves, shaded leaves, and soil (both sunlit and shaded).

If the BRF for soils and vegetation were isotropic; that is, the surfaces responded like Lambertian surfaces, then the magnitude of the BRF would be constant for all view angles. However the BRF for canopies can vary by more than a factor of three with view angle for a given wavelength band. Detailed models of canopy BRFs are complex and beyond the scope of this book. Even analytical models such as Kuusk (1995) are quite complicated. However, Irons et al. (1992) have represented the soil BRF by small spheres on a flat Lambertian plane, where the shadows cast by the spheres onto the horizontal background influence the radiation viewed by the sensor. The BRF distributions for canopies and soils have a characteristic shape with BRF values being highest when the sun is directly behind the sensor and low when the sensor view is directed toward the sun. Walthall et al. (1985) present a simple, empirical equation to fit BRF distributions as a function of view zenith and view azimuth for a single sun zenith angle:

$$\text{BRF} = a\psi_v^2 + b\psi_v \cos(\Delta AZ) + c \qquad (15.25)$$

where ψ_V is the view zenith angle, ΔAZ is the difference between the azimuth angle of the sensor and the azimuth angle of the sun ($\Delta AZ = 0$ when the sun is directly behind the viewer so the view is away from the direction of the sun), and a, b, and c are empirical coefficients that change with canopy architecture, wavelength, and sun zenith angle. For example, Walthall et al. (1985) give the coefficients for a soybean canopy with LAI= 2.6 and $\psi = 61°$ as VIS $a = 1.49$, $b = 0.32$, and $c = 3.44$, with NIR $a = 9.09$, $b = 7.62$, and $c = 46.8$ (BRF in % and angles in radians). Clearly the BRF is largest when the middle term of Eq. (15.25) is positive and smallest when the middle term is negative.

In this chapter we are interested in understanding the relation between BRF and canopy architecture: This can be accomplished with simplified equations by limiting the discussion to a sensor viewing from near nadir (or within about 10° of directly overhead). This is the most common direction used in remote sensing because atmospheric contamination is minimal and interpretation of nadir data is most straightforward. In the following sections, ψ refers to the sun zenith angle, and since the sun zenith angle is rarely zero, we use 0 to refer to the nadir view angle so that $K_{be}(\psi)$ refers to the extinction coefficient for beam radiation and $K_{be}(0)$ refers to the extinction from the nadir view direction. $K_{be}(0)$ is never used in the following equations to refer to the direction of the sun. Although similar equations are used to describe sun and view effects, the context should always be obvious.

From Eq. (15.22), the fraction of leaves at depth L in a canopy that is sunlit is given by $\exp(-K_{be}(\psi)L)$. If the sensor is placed at the same zenith and azimuth angles as the sun, and for discussion purposes assume the sensor is so small that it casts a negligibly small shadow, then the sensor would view exactly these sunlit leaves. Thus the same exponential expression can be used to estimate sunlit-leaf-area fraction as to estimate the fraction of leaves in a layer than can be viewed from the same direction. The special case of identical sun and view directions is referred to as the canopy hot spot, because the canopy appears brighter from this direction then any other direction; all the leaves being sunlit from this direction. For this special case, sunlit leaves and shaded leaves cannot be assumed to be independent because both sun and view directions share the same path through the canopy. As the sensor is moved off the direction of the sun, shaded leaves occupy an increasing fraction of the sensor FOV until the view path and the path of the sun rays are independent. The decrease in radiance as a function of the increasing angle between the direction of the sensor and the sun direction depends on LAI, leaf angle distribution, leaf size, sun zenith angle, and canopy height (Kuusk, 1995). Typically the hot spot varies from a few degrees wide to a few tens of degrees wide depending on conditions. The equations that follow, which pertain to nadir viewing only, do not consider the hot spot. Hot spot considerations generally would represent a minor refinement for latitudes where the sun zenith angle is rarely less than 20°.

A remote sensor that is directed toward a canopy–soil system may view both vegetation and soil. For a canopy of randomly positioned leaves, the fraction of the sensor view that is occupied by soil is $\exp(-K_{be}(0)L_t)$ so that the fraction of view occupied by vegetation is $1 - \exp(-K_{be}(0)L_t)$. The vegetation portion of the view consists of both sunlit and shaded leaves. The fraction of leaves at a depth L that is sunlit with the sun at zenith angle ψ is given by Eq. (15.22). Therefore the fraction of leaves at a depth L that is sunlit and can be viewed from nadir is given by the product: $\exp(-K_{be}(0)L) \exp(-K_{be}(\psi)L)$. If the product of these two exponentials is integrated over the depth of the canopy L_t the sunlit leaf area index is obtained that is in the view of the sensor, L_V^*:

$$L_V^* = \frac{1 - \exp\left\{ - \left[K_{be}(0) + K_{be}(\psi)\right] L_t\right\}}{K_{be}(0) + K_{be}(\psi)}. \tag{15.26}$$

The fraction of the sensor view occupied by sunlit leaves is given by the projection of L_V^* in the direction of the sensor or $f_{V,sl} = L_V^* K_{be}(0)$. The fraction of sensor view occupied by shaded leaves is the difference between the view fraction occupied by vegetation $(1 - \exp(-K_{be}(0)L_t))$

and the fraction occupied by sunlit leaves:

$$f_{V,sh} = 1 - \exp(-K_{be}(0)L_t) - L_V^* K_{be}(0). \qquad (15.27)$$

The flux density detected by the sensor, $Q_{view}(\psi)$, is the sum of the contributions of sunlit leaves, shaded leaves, and soil weighted by the view fractions each occupies:

$$\begin{aligned} Q_{view}(\psi) = {}& \rho \overline{Q_{sl}(\psi)} L_V^* K_{be}(0) \\ &+ \rho \overline{Q_{sh}} \left[1 - \exp(-K_{be}(0)L_t) - L_V^* K_{be}(0)\right] \quad (15.28) \\ &+ \rho_s (Q_{bt} + Q_d) \exp(-K_{be}(0)L_t) \end{aligned}$$

where ρ and ρ_s are the leaf and soil reflectivity in the wavelength band of interest. The BRF for a particular wavelength band, for example, the visible, is given by

$$BRF_V = \left(\frac{Q_{view}(\psi)}{Q_{ob} + Q_{od}}\right)_V. \qquad (15.29)$$

The unique feature of leaves that permits remote sensing of canopy bidirectional reflectance to be useful for estimating canopy biophysical characteristics is the strong contrast between absorption in the visible and scattering in the near-infrared with a sharp transition near 700 nm (Fig. 11.5). Usually soils have higher reflectivity in the visible than dense canopies, lower reflectivities in the near-infrared than dense canopies, and only slightly higher reflectivity in the near-infrared than visible; therefore as canopy cover increases, the visible reflectance decreases, near-infrared reflectance increases, and the ratio, given by

$$SR = \frac{BRF_N}{BRF_V} \qquad (15.30)$$

increases (SR is called the simple ratio vegetation index). Another form of the ratio is the normalized difference vegetation index (NDVI) given by

$$NDVI = \frac{BRF_N - BRF_V}{BRF_N + BRF_V} \qquad (15.31)$$

where $-1 \le NDVI \le 1$. These vegetation indices in the form of ratios are widely used in remote sensing because uncertainties that affect both wavelength bands similarly tend to cancel out. Numerous other indices have been developed to minimize the influence of atmospheric or soil contamination and the advantage gained from these variations over SR and NDVI appears to be minor but consistent. NDVI may not be zero for zero vegetation cover because soil reflectances in the two bands may not be equal or because of atmospheric effects (VIS is scattered more than NIR so NDVI can be negative from satellite observations if no atmospheric corrections are done); therefore, an adjusted NDVI (NDVI*) has been proposed by Carlson et al. (1995):

$$NDVI^* = \left(\frac{NDVI - NDVI_{min}}{NDVI_{max} - NDVI_{min}}\right)^2 \qquad (15.32)$$

where $NDVI_{min}$ is the NDVI with no vegetation and $NDVI_{max}$ is the NDVI with dense vegetation. Carlson, et al. (1995) set NDVI* equal to the fraction of vegetative cover, and this is a reasonable approximation, especially when solar zenith angles are small. Clearly NDVI* varies from zero to one over the range of vegetation cover and accounts for the observation that NDVI increases more rapidly than the fraction of vegetation cover as vegetation density increases.

Remote sensing from satellites has the possibility of sampling the entire land surface of the earth daily at a 1 km spatial resolution on the ground and a spatial resolution of 10 m, or less with less frequent temporal sampling. Because of this phenomenal spatial sampling, much effort has been expended to determine what biophysical quantities are most closely related to the remote sensing observations. An examination of Eq. (15.28) provides some useful insights here. Remember that optical remote sensing from satellites is possible only under relatively clear-sky conditions when atmospheric transparency is high, because satellites need to view the surface with minimal contamination from the atmosphere. From Eq. (15.28), when L_t is small, reflection from the soil dominates (third term on the right of Eq. (15.28)). As L_t increases, the dominant term in Eq. (15.28) becomes the scattering of intercepted near-infrared beam radiation (first term on the right of Eq. (15.28)), which also happens to be closely related to the intercepted PAR radiation. Table 15.2 contains values of the three terms in Eq. (15.28), NDVI, NDVI*, and IPAR and the fraction of canopy cover (f_c) assuming

$$f_c = \exp(-K_{be}(0)L_t). \tag{15.33}$$

Clearly NDVI* is most closely related to IPAR and fraction vegetative cover (f_c) when ψ is small (30°). The relation between NDVI* and IPAR is likely to be better at other solar zenith angles because both NDVI* and IPAR change but f_c is fixed with ψ. The close relation between NDVI* and IPAR occurs because intercepted solar radiation dominates both variables; interception in the visible portion of the solar spectrum dominates IPAR and interception in the NIR portion of the solar spectrum dominates NDVI*.

The effects of leaf angle and sun zenith angle can be seen from Table 15.3. Clearly NDVI* is a reasonable predictor of fraction of IPAR for a modest range of conditions. Since IPAR is closely related to vegetation productivity potential (Eq. (14.13) with S_t replaced by IPAR and conversion efficiency e adjusted accordingly [e is about doubled]), remote sensing has something significant to contribute to global vegetation studies. The robustness of the relation between NDVI* and fraction of IPAR is further established by studies that have shown NDVI* to be related to the fraction of IPAR associated with the green vegetation in canopies that have both green and dead foliage.

Example 15.3. Compare the nadir, near-infrared BRF (BRF_N) for a canopy with a spherical leaf angle distribution ($x = 1$) with the hemi-

TABLE 15.2. Variation of some quantities related to remote sensing as a function of several canopy biophysical characteristics. The wavelengths used for remote sensing calculations are about 650 nm and 750 nm. The three terms from Eq. (15.28) are for the NIR wavelength band. The canopy is assumed to have a spherical leaf angle distribution and $\psi = 30°$. All fluxes are in units of W m^{-2}.

L_t	NIR Term 1 (W m^{-2})	NIR Term 2 (W m^{-2})	NIR Term 3 (W m^{-2})	NDVI	NDVI*	IPAR (W m^{-2})	Fraction IPAR	Fraction Cover
0	0	0	75	0.20	0	0	0	0
0.1	9	0	70	0.27	0.01	28	0.06	0.06
0.3	25	1	61	0.39	0.06	78	0.16	0.14
0.6	47	3	50	0.54	0.20	143	0.29	0.26
1.0	68	9	38	0.68	0.41	212	0.42	0.39
1.5	86	17	28	0.78	0.59	280	0.56	0.53
2.0	98	25	20	0.84	0.72	331	0.66	0.63
4.0	113	50	6	0.91	0.90	441	0.88	0.86
6.0	113	57	2	0.92	0.92	479	0.96	0.95

spherical near-infrared reflectance for a sun zenith angle $\psi = 60°$, Assume a leaf reflectivity and transmissivity of 0.48 so $\alpha_N = 0.04$, soil reflectance $\rho_s = 0.15$, and $L_t = 2.0$. The near-infrared part of the incident solar radiation is $Q_{ob} = 230$ W m^{-2} and $Q_{od} = 20$ W m^{-2}.

Solution. The canopy BRF$_N$ is estimated from Eq. (15.28) so the three terms in that equation need to be evaluated. The following quantities are

TABLE 15.3. Relation between NDVI* and IPAR fraction (f_{IPAR}) for two sun zenith angles and two leaf angle distributions.

L_t	$x = 1$				$x = 4$			
	$\psi = 30$		$\psi = 60$		$\psi = 30$		$\psi = 60$	
	NDVI*	f_{IPAR}	NDVI*	f_{IPAR}	NDVI*	f_{IPAR}	NDVI*	f_{IPAR}
0	0		0		0		0	
0.1	0.01	0.06	0.02	0.08	0.03	0.08	0.03	0.08
0.3	0.06	0.16	0.11	0.23	0.18	0.21	0.19	0.22
0.6	0.20	0.29	0.31	0.40	0.42	0.38	0.43	0.40
1.0	0.41	0.42	0.52	0.58	0.62	0.54	0.64	0.56
1.5	0.59	0.56	0.71	0.72	0.75	0.69	0.77	0.71
2.0	0.72	0.66	0.80	0.81	0.82	0.79	0.82	0.80
4.0	0.90	0.88	0.90	0.96	0.87	0.95	0.87	0.96
6.0	0.92	0.96	0.91	0.99	0.89	0.99	0.89	0.99

needed:

$$K_{be}(0) = \frac{\sqrt{1.0 + \tan^2(0)}}{1.0 + 1.774(1.0 + 1.182)^{-0.733}} = 0.50$$

$$K_{be}(60) = \frac{\sqrt{1.0 + \tan^2(60)}}{1.0 + 1.774(1.0 + 1.182)^{-0.733}} = 1.00$$

$$K_d = 0.76 \qquad \text{(from Fig. 15.4)}$$

$$L_V^* = \frac{1 - \exp\left[-(0.5 + 1.0)2.0\right]}{0.5 + 1.0} = 0.633$$

$$Q_{\text{view}}(\psi) = \text{Term 1} + \text{Term 2} + \text{Term 3}$$

$$\text{Term 1} = \rho \overline{Q_{sl}}(\psi) L_V^* K_{be}(0)$$

$$\overline{Q_{sl}}(60) = K_{be}(60) Q_{ob} + \overline{Q_d} + \overline{Q_{sc}}$$

$$Q_d = 20 \exp(-\sqrt{0.04} \times 0.76 \times 2.0) = 20(0.738)$$

$$= 15 \text{ W m}^{-2}$$

$$\overline{Q_d} = \frac{(20 + 15)}{2} = 18 \text{ W m}^{-2}$$

$$Q_{sc}(60) = 230 \left[\exp(-\sqrt{0.04} \times 1.0 \times 2.0) - \exp(-1.0 \times 2.0)\right]$$

$$= 123 \text{ W m}^{-2}$$

$$\overline{Q_{sc}} = \frac{0 + 123}{2} = 62 \text{ W m}^{-2}$$

$$\overline{Q_{sl}}(60) = 1.0(230) + 18 + 62 = 310 \text{ W m}^{-2}$$

$$\text{Term 1} = 0.48 \times 310 \times 0.633 \times 0.5 = 47 \text{ W m}^{-2}$$

$$\text{Term 2} = \rho \overline{Q_{sh}} \left[1 - \exp(-K_{be}(0)L_t) - L_V^* K_{be}(0)\right]$$

$$\overline{Q_{sh}} = \overline{Q_d} + \overline{Q_{sc}} = 18 + 62 = 80 \text{ W m}^{-2}$$

$$\text{Term 2} = 0.48 \times 80 \left[1 - \exp(-0.5 \times 2.0) - 0.633 \times 0.5\right]$$

$$= 12 \text{ W m}^{-2}$$

$$\text{Term 3} = \rho_s(Q_{bt} + Q_d) \exp(-K_{be}(0)L_t)$$

$$Q_{bt}(60) = 230 \exp\left[-\sqrt{0.04} \times 1.0 \times 2.0\right] = 230 \times 0.67$$

$$= 154 \text{ W m}^{-2}$$

$$\text{Term 3} = 0.15 \times (154 + 15) \exp(-0.5 \times 2.0) = 9 \text{ W m}^{-2}.$$

Therefore $Q_{\text{view}}(60) = 47 + 12 + 9 = 68 \text{ W m}^{-2}$ so that $\text{BRF}_N = \frac{68}{230+20} = 0.27$. If we had used the more precise Eq. (15.11) instead of Eq. (15.6), then $\text{BRF}_N = 0.32$ instead of 0.27.

The hemispherical reflectance can be estimated from Eq. (15.10) for both beam and diffuse components. For the beam component:

$$\rho_{b,\text{cpy}}(60) \simeq \rho^*_{b,\text{cpy}}(60) - (\rho^*_{b,\text{cpy}}(60) - \rho_s) \exp(-2\sqrt{\alpha}K_{be}(60)L_t)$$

$$\rho^H_{\text{cpy}} = \frac{1 - \sqrt{0.04}}{1 + \sqrt{0.04}} = \frac{0.8}{1.2} = 0.667 \qquad \text{(Eq. (15.7))}$$

$$\rho^*_{b,\text{cpy}} = \frac{2 \times 1.0}{1.0 + 1.0}(0.667) = 0.667 \qquad \text{(Eq. (15.8))}$$

$$\rho_{b,\text{cpy}}(60) = 0.667 - (0.667 - 0.15)\exp(-2 \times \sqrt{0.04} \times 1.0 \times 2.0)$$
$$= 0.435$$

$$\rho_{d,\text{cpy}} \simeq \rho^*_{d,\text{cpy}} - (\rho^*_{d,\text{cpy}} - \rho_s)\exp(-2\sqrt{\alpha}K_dL_t)$$

$$\rho^*_{d,\text{cpy}} = \frac{2.0 \times 0.76}{0.76 + 1}(0.667) = 0.576$$

$$\rho_{d,\text{cpy}} = 0.576 - (0.576 - 0.15)\exp(-2 \times \sqrt{0.04} \times 0.76 \times 2.0)$$
$$= 0.344.$$

Therefore the hemispherical reflectance of the canopy is

$$\rho_{\text{cpy}}(60) = \frac{Q_{ob}\rho_{b,\text{cpy}}(60) + Q_{od}\rho_{d,\text{cpy}}}{Q_{ob} + Q_{od}}$$
$$= \frac{230 \times 0.435 + 20 \times 0.344}{250} = 0.43.$$

If we had used the more precise Eq. (15.9) instead of Eq. (15.10), then $\rho_{\text{cpy}}(60) = 0.48$ instead of 0.43.

The reason BRF_N is lower than $\rho_{\text{cpy}}(60)$ in the near-infrared is that the nadir-viewing sensor views deeper into the canopy than the sun penetrates and thus the nadir BRF_N is lower by 37 percent. This indicates the undesirability of using hemispherical reflectances to make inferences about remote sensing with narrow FOV sensors.

15.11 Remote Sensing and Canopy Temperature

Aerodynamic surface temperature is a key variable in the partitioning of net radiation into sensible and latent heat fluxes, as shown in Ch. 14, particularly in Eq. (14.8). Since radiometric surface temperature is a quantity that can be measured from satellites over the globe on kilometer spatial scales, numerous attempts have been made to use these remotely-sensed radiometric temperatures to monitor the partitioning of sensible and latent heat fluxes. The magnitude of this challenge is apparent from examining Eq. (14.8); obviously many variables can affect aerodynamic surface temperature, and the additional variables involved in the relation between radiometric and aerodynamic temperatures are not even included in Eq. (14.8). Although radiometric temperature may be available globally, most of the other variables that affect surface temperature are not.

The sensible heat flux from the vegetation/soil system is closely related to surface aerodynamic temperature by

$$H_{cpy} = c_p g_{Ha}(T_{aero} - T_a) \qquad (15.34)$$

where g_{Ha} is the aerodynamic conductance or canopy boundary-layer conductance given by Eq. (14.9) and T_a is the air temperature. The apparent simplicity of Eq. (15.34) is deceptive. Assuming the information is available on a continental basis to estimate g_{Ha}, and this is no minor task because vegetation height, cover, and wind speed are required (remote sensing of NDVI may help here), three major challenges remain in trying to use radiometric temperature to estimate sensible heat flux:

1. The radiometric temperature and aerodynamic temperature are not the same and usually differ by 1 to 5° C.
2. The near-surface air temperature is not known on the same spatial scale as radiometric temperature and can vary by 5° C or more depending on the temperature of the underlying surface.
3. Atmospheric corrections and uncertainties in surface emissivity associated with satellite-borne surface radiometric temperatures have uncertainties of 1 to 3° C.

Unfortunately, an uncertainty of 1° C in $T_{aero} - T_a$ can result in a 50 W m^{-2} uncertainty in H_{cpy}; a reasonable estimate of a tolerable maximum error. These challenges have not deterred scientists from searching for a solution.

From this discussion a practical method for using satellite surface temperature measurements should have at least three qualities:

1. Accommodate the difference between aerodynamic temperature and radiometric temperature.
2. Not require a measurement of near-surface air temperature.
3. Rely more on differences of surface temperature over time or space rather than absolute surface temperatures to minimize the influence of atmospheric corrections and uncertainties in surface emissivity.

Anderson et al. (1997) have proposed such a method based on satellite observations from the Geosynchronous Orbiting Environmental Satellite (GOES), which is used primarily for observations of clouds and weather forecasting, having a ground spatial resolution of 4 km. In addition to the satellite temperature observations, they use ground measurements and balloon measurements from the weather forecasting network, a continental vegetation classification map, and vegetation cover estimated with NDVI as described in the previous section. Uncertainties in sensible and latent heat of 30 to 50 W m^{-2} are achievable by this method. Practical methods for using satellite observations of surface temperature to partition sensible and latent heat fluxes on a continental scale are most challenging.

15.12 Canopy Reflectivity (Emissivity) versus Leaf Reflectivity (Emissivity)

Canopy reflectance is less than leaf reflectance because some of the radiation incident on leaves is transmitted deeper into the canopy where multiple interactions between the radiation and leaves causes additional absorption of the radiation. In effect, the canopy behaves as a trap for the radiation that is absorbed at the deeper depths in the canopy or at the soil surface. Either Eq. (15.7) or Eq. (15.8) can be used to illustrate this trapping phenomenon. For a deep canopy with PAR reflectivity $\rho_p = 0.1$ and PAR transmissivity $\tau_p = 0.1$, the canopy reflectance $\rho_{cpy}^H = 0.056$. In the thermal wavelength band, if the leaf emissivity $\varepsilon_L = 0.95$, then the leaf reflectivity $\rho_L = 0.05$ because $\tau_L = 0$. Using Eq. (15.7) for a deep canopy, $\rho_{L,cpy} = 0.013$ so the emissivity of this deep canopy is 0.987. Therefore a deep canopy is much closer to a blackbody than the leaves that make it up, and this explains why dense canopies often are assumed to have thermal emissivities of 0.99 even though leaves may have lower emissivities.

15.13 Heterogeneous Canopies

The simplified radiative exchange principles described in this chapter apply to vegetative canopies with leaves that are randomly distributed throughout the canopy space. Such canopies of randomly-positioned leaves are often referred to as homogeneous because the probability of finding a leaf anywhere in the canopy space is independent of horizontal position. When leaves are not randomly distributed in space, the canopy is considered heterogeneous; and the character of the heterogeneity can take many forms. We briefly consider two approaches to characterizing heterogeneity.

1. Incorporate a clumping factor in the exponential extinction equations by replacing L with $\Omega(\psi)L$; where $\Omega(\psi)$ is the clumping factor that depends on zenith angle.
2. Assume leaves to be randomly distributed within the confines of some appropriate geometric volumes, which we refer to as canopy envelopes, to represent widely-spaced tree crowns or crop rows.

The clumping-factor approach has the advantage of making it possible to extend the previous equations for random canopies discussed earlier in this chapter to heterogeneous cases. For random canopies $\Omega(\psi) = 1$, clumped foliage has $\Omega(\psi) < 1$, and if foliage is more nearly uniformly spaced, $\Omega(\psi) > 1$. For forest canopies, which tend to be the most strongly clumped, the dependence of clumping factor on ψ can be approximated by the following equations:

$$\Omega(\psi) = \frac{\Omega(0)}{\Omega(0) + [1 - \Omega(0)] \exp[-2.2(\psi)^p]}$$

$$p = 3.80 - 0.46D \quad 1 \le p \le 3.34 \tag{15.35}$$

$$D = \frac{\text{crown depth}}{\text{crown diameter}}$$

where $\Omega(0)$ is the clumping factor when the canopy is viewed from nadir or when looking up out of the canopy toward the zenith. Table 15.4 contains some values of $\Omega(0)$ for mature stands of several species.

Using Eq. (15.35), sunlit leaf area index can be estimated for a clumped canopy by using Eq. (15.23) and replacing L_t by $\Omega(\psi)L_t$, and diffuse penetration estimated from the same substitution into Eq. (15.5). This approach is only approximate because the scattering equations imply a random distribution of leaves.

With conifers, an additional level of clumping occurs because needles are organized onto shoots. Typically the hemi-surface area of conifer shoots is about 1.3 to 2 times greater than the effective light-intercepting area of shoots. This shoot clumping factor is quite important when canopy architecture is estimated from indirect measurements such as those discussed in the next section. Fassnacht et al. (1994) describe a method for estimating shoot clumping factors, and show that the difference in HSAI of fertilized and unfertilized pine stands is 30 percent; with 23 percent of this difference arising because fertilized shoots contain more needle surface area (more strongly clumped) and only seven percent difference arising from the increased light interception as determined by an indirect measurement of HSAI.

The second approach to characterizing heterogeneous canopies requires knowing the dimensions of geometric canopy envelopes that contain all the foliage. This approach is most useful when the spatial distribution of radiation beneath canopies is needed; such as in agroforestry where crop placement beneath tree crowns may be critical. If canopy envelopes are assumed to be ellipsoids, such as Norman and Welles (1983) use, then a wide variety of crown shapes can be simulated. Given an ar-

TABLE 15.4.   Canopy clumping factors in the zenith direction for mature, healthy stands of several species.

Species (Location)	Hemi-Surface Area Index	D	$\Omega(0)$
Sugar Maple (Northern Wisconsin, U.S.A.)	5.5	~ 1	0.95
Oak (North Carolina, U.S.A.)	4	~ 1	0.9
Aspen (Saskatchewan, Canada)	3.5	1.5–2	0.7
Jack Pine (Saskatchewan, Canada)	2.5	3–4	0.5
Black Spruce (Saskatchewan, Canada)	6.5	5–6	0.4

ray of canopy envelopes of known dimensions and locations, the beam transmittance $\tau_b(\psi, AZ)$ can be estimated from

$$\tau_b(\psi, AZ) = \exp(-K_{be}(\psi)\mu S(\psi, AZ) \cos \psi) \qquad (15.36)$$

where μ is the leaf area density (m^2 of hemi-surface area per m^3 canopy volume) and $S(\psi, AZ)$ is the path length of light rays through the array of canopy envelopes between a particular point in a horizontal (at some depth in the canopy or at the soil surface) plane and the sun.

Models of BRF in heterogeneous canopies are quite complicated and several approaches are described in detail in a book edited by Myneni and Ross (1991).

15.14 Indirect Sensing of Canopy Architecture

A description of canopy architecture includes the position and orientation distributions of leaves, branches, stems, flowers, and fruit. For most canopies, leaves dominate the canopy space so leaf area index, leaf angle distribution and some measure of clumping provide most of the information needed to describe canopy architecture. If we limit our discussion to canopies that approximate random positioning (most full-cover deciduous forests, grasslands and crops), then LAI and x are the minimum essential bits of information. Direct measurements of LAI and x, by cutting plants and measuring leaf areas and angles are exceedingly laborious, so alternative measurement methods are desirable. Measurements of canopy gap fraction as a function of zenith angle can be used to obtain estimates of L_t and $K_{be}(\psi)$. The strategy for using gap-fraction measurements to estimate canopy architecture is illustrated in Fig. 15.3. The gap fraction corresponds to the ordinate labeled transmission and the curves show the effect of leaf angle distribution (x) on transmission or gap fraction as a function of zenith angle for $L_t = 1$. Given a number of measurements of gap fraction as a function of zenith angle, the curve that best fits the data can be chosen from numerous families of curves such as shown in Fig. 15.3 calculated for a range of LAI values. The values of x and L_t that best fit the data are assigned to the canopy where the gap-fraction measurements originated (Norman and Campbell, 1989). Although this method appears to be simple, the inversion procedure can be error prone and must be done carefully. Several commercial instruments that use this approach are available and have been discussed by Welles (1990).

Heterogeneous (nonrandom) canopies require some additional information about the characteristics of the heterogeneity. If canopy heterogeneity can be represented by the parameter $\Omega(\psi)$ in Eq. (15.35), then additional methods must be available for estimating $\Omega(\psi)$ (Chen, 1996) beyond the measurements of gap fraction as a function of zenith angle.

If heterogeneous canopies are composed of regular geometric shapes that contain foliage with large gaps between them, then the path length $S(\psi, AZ)$ may be determined for the particular geometry (horizontal cylinders for row crops or regularly spaced spheres for an orchard) and added to the inversion process (Welles, 1990).

Indirect methods exist to estimate L_t and x for a wide variety of homogeneous and heterogeneous canopies including prairies, row crops, deciduous, and coniferous forests. Even though direct destructive measurements remain the reference standards for evaluating the accuracy of indirect methods, indirect measurements are faster, easier, and provide better spatial sampling.

The indirect sensing of canopy architecture provides an example of how an improved understanding of the fundamentals of radiative exchange in vegetation has provided a solution to the practical problem of characterizing plant canopies.

References

Anderson, M. C., J.M. Norman, G.R. Diak, W.P. Kustas, and J.R. Mecikalski (1997) A two-source time-integrated model for estimating surface fluxes using thermal infrared remote sensing. Remote Sensing of Environment 60:195–216.

Campbell, G. S. (1986) Extinction coefficients for radiation in plant canopies calculated using an ellipsoidal inclination angle distribution. Agricultural and Forest Meteorology 36:317–321.

Campbell, G.S. (1990) Derivation of an angle density function for canopies with ellipsoidal leaf angle distributions. Agricultural and Forest Meteorology 49:173–176

Campbell, G. S. and F. K. van Evert (1994) Light interception by plant canopies: efficiency and architecture. In J. L. Monteith, R. K. Scott, and M. H. Unsworth, Resource Capture by Crops. Nottingham University Press.

Carlson, T,N., W.J. Capehart, and R.R. Gillies (1995) A new look at the simplified method for remote sensing of daily evapotranspiration. Remote Sensing of Environment 54:161–167.

Chen, J.M. 1996. Optically-based methods for measuring seasonal variation of leaf area index in boreal conifer stands. Agricultural and Forest Meteorology 80:135–163.

Fassnacht, K.S., S.T. Gower, J.M. Norman, and R.E. McMurtrie (1994) A comparison of optical and direct methods for estimating foliage surface area index in forests. Agricultural and Forest Meteorology 71:183–207.

Fuchs. M., G. Stanhill, and S. Moreshet (1976) Effect of increasing foliage and soil reflectivity on the solar radiation balance of wide-row sorghum. Agronomy Journal 68:865–871.

Goudriaan, J. (1977) Crop micrometeorology: a simulation study. Simulation Monographs, Pudoc, Wageningen.

Goudriaan, J. (1988) The bare bones of leaf-angle distribution in radiation models for canopy photosynthesis and energy exchange. Agricultural and Forest Meteorology 43:155–169.

Irons, J., G.S. Campbell, J.M. Norman, D.W. Graham, and W.K. Kovalick (1992) Prediction and measurement of soil bidirectional reflectance. IEEE Transactions of Geoscience and Remote Sensing. 30:249–260.

Kuusk, A. (1991) The hot spot effect in plant canopy reflectance. In Photon-Vegetation Interactions: Applications in Optical Remote Sensing and Plant Ecology. Eds. R.B. Myneni and J. Ross. Springer-Verlag, Berlin. pp.139–159.

Kuusk, A. (1995) A fast, invertible canopy reflectance model. Remote Sensing of Environment. 51:342–350.

Monteith, J.L. and M.H. Unsworth. (1990) Principles of Environmental Physics. Edward Arnold Publishers, London. 291 pp.

Myneni, R.B. and J. Ross. (eds.) (1991) Photon-Vegetation Interactions: Applications in Optical Remote Sensing and Plant Ecology. Springer-Verlag, Berlin. 565 pp.

Norman, J.M. (1992) Scaling processes between leaf and canopy levels. In Scaling Physiological Processes: Leaf to Globe. Eds. J.R. Ehleringer and C.B. Field. Academic Press, Inc. San Diego, CA pp. 41–76.

Norman, J.M. and G.S. Campbell (1989) Canopy structure. In Plant Physiological Ecology: Field Methods and Instrumentation. Eds. R.W. Pearcy, J. Ehleringer, H.A. Mooney, and P.W. Rundel. Chapman and Hall, N.Y. pp. 301–325.

Norman, J.M. and J.M. Welles (1983) Radiative transfer in an array of canopies. Agronomy Journal 75:481–488.

Walthall, C.L., J.M. Norman, J.M. Welles, G.S. Campbell, and B.L. Blad (1985) Simple equation to approximate the bidirectional reflectance from vegetative canopies and bare soil surfaces. Applied Optics. 24:383–387.

Welles, J.M. (1990) Some indirect methods for determining canopy structure. Remote Sensing Reviews. 5:31–43.

Problems

15.1. A canopy with a spherical leaf angle distribution has a total leaf area index of three. Find the flux density of PAR on sunlit and on shaded leaves at the bottom of the canopy, and the fraction of the leaves which are sunlit and shaded. Assume a clear sky with a solar zenith angle of 30°.

15.2. Find the daily fractional transmission of PAR, NIR, and total solar radiation by a canopy with leaf area index, $L_t = 2$. Assume that the leaf angle distribution is approximated by an ellipsoidal angle distribution with $x = 2$.

15.3. If the ratio of red to far red radiation at the top of the canopy in problem 15.1 is 1, what is the ratio at the bottom of the canopy. Assume $\alpha_{red} = 0.8$ and $\alpha_{far\ red} = 0.2$.

15.4. Using Eq. (15.25) and the coefficient values in the text for VIS and NIR wavelength bands, plot the BRF_N and BRF_V as a function of view zenith angle for the principal plane of the sun between nadir and 60°. The principal plane occurs when $\Delta AZ = 0$ or $\Delta AZ = \pi$. The horizontal axis of the graph will go from zenith view angles of $-60°$ to $+60°$ with positive view angles corresponding to $\Delta AZ = 0$ and negative zenith view angles corresponding to $\Delta AZ = \pi$. (with LAI $= 2.6$ and $\psi = 61°$): for VIS $a = 1.49$, $b = 0.32$, and $c = 3.44$, and NIR $a = 9.09$, $b = 7.62$, and $c = 46.8$ (BRF in % and angles in radians). Considering that $\Delta AZ = 0$ corresponds to having the sun behind the sensor and $\Delta AZ = \pi$ corresponds to the viewer looking toward the sun but downward at the canopy, explain the shape of this curve.

Appendix

TABLE A1. Temperature dependent properties of gases in air at 101 kPa.

T C	ρ mol m^{-3}	ν mm^2/s	D_H mm^2/s	D_v mm^2/s	D_c mm^2/s	D_o mm^2/s
0	44.6	13.3	18.9	21.2	13.9	17.7
5	43.8	13.7	19.5	21.9	14.3	18.3
10	43.0	14.2	20.1	22.6	14.8	18.8
15	42.3	14.6	20.8	23.3	15.3	19.4
20	41.6	15.1	21.4	24.0	15.7	20.0
25	40.9	15.5	22.0	24.7	16.2	20.6
30	40.2	16.0	22.7	25.4	16.7	21.2
35	39.5	16.4	23.3	26.2	17.2	21.9
40	38.9	16.9	24.0	26.9	17.7	22.5
45	38.3	17.4	24.7	27.7	18.2	23.1

Specific heat of air: $c_p = 29.3$ J mol^{-1} C^{-1}
Molecular mass of air: $M_a = 29$ g/mol.
Molecular mass of water: $M_w = 18$ g/mol.

TABLE A2. Properties of water

T C	ρ_w MG/m^3	λ kJ/mol	ν mm^2/s	D_H mm^2/s	D_o mm^2/s	D_v mm^2/s
0	0.99987	45.0	1.79	0.134		
4	1.00000	44.8	1.57	0.136		
10	0.99973	44.6	1.31	0.140		
20	0.99823	44.1	1.01	0.144	0.002	0.002
30	0.99568	43.7	0.80	0.148		
40	0.99225	43.4	0.66	0.151		
50	0.98807	42.8	0.56	0.154		

Specific heat of water   75.4 J mol^{-1} C^{-1}
Latent heat of freezing   6.0 kJ mol^{-1}
Thermodynamic psycrometer constant at 20 C   0.000664 C^{-1}

TABLE A3. Temperature dependence of saturation vapor pressure,
slope of the vapor pressure function, black body emittance, radiative
conductance, and clear sky emissivity.

Temp K	Temp C	$e_s(T)$ kPa	Δ PaC^{-1}	B W m^{-2}	g_r mol m^{-2} s^{-1}	ε_α
268.2	-5	0.422	32	293	0.149	0.66
269.2	-4	0.455	34	298	0.151	0.67
270.2	-3	0.490	36	302	0.153	0.67
271.2	-2	0.528	39	307	0.154	0.68
272.2	-1	0.568	42	311	0.156	0.68
273.2	0	0.611	44	316	0.158	0.69
274.2	1	0.657	47	320	0.160	0.69
275.2	2	0.706	50	325	0.161	0.70
276.2	3	0.758	54	330	0.163	0.70
277.2	4	0.813	57	335	0.165	0.71
278.2	5	0.872	61	339	0.167	0.71
279.2	6	0.935	65	344	0.168	0.72
280.2	7	1.001	69	349	0.170	0.72
281.2	8	1.072	73	354	0.172	0.73
282.2	9	1.147	77	359	0.174	0.73
283.2	10	1.227	82	365	0.176	0.74
284.2	11	1.312	87	370	0.178	0.74
285.2	12	1.402	92	375	0.179	0.75
286.2	13	1.497	98	380	0.181	0.75
287.2	14	1.597	104	386	0.183	0.76
288.2	15	1.704	110	391	0.185	0.76
289.2	16	1.817	116	396	0.187	0.77
290.2	17	1.936	123	402	0.189	0.77
291.2	18	2.062	130	407	0.191	0.78
292.2	19	2.196	137	413	0.193	0.79
293.2	20	2.336	145	419	0.195	0.79
294.2	21	2.485	153	425	0.197	0.80
295.2	22	2.642	161	430	0.199	0.80
296.2	23	2.808	170	436	0.201	0.81
297.2	24	2.982	179	442	0.203	0.81
298.2	25	3.166	189	448	0.205	0.82
299.2	26	3.360	199	454	0.207	0.82
300.2	27	3.564	209	460	0.209	0.83
301.2	28	3.778	220	466	0.211	0.83
302.2	29	4.004	232	473	0.214	0.84
303.2	30	4.242	244	479	0.216	0.85
304.2	31	4.492	256	485	0.218	0.85
305.2	32	4.754	269	492	0.220	0.86
306.2	33	5.030	283	498	0.222	0.86
307.2	34	5.320	297	505	0.224	0.87
308.2	35	5.624	311	511	0.227	0.87
309.2	36	5.943	327	518	0.229	0.88
310.2	37	6.278	343	525	0.231	0.89

Continued on next page

TABLE A3.   (*continued*)

Temp K	Temp C	$e_s(T)$ kPa	Δ Pa	B W m^{-2}	g_r mol m^{-2} s^{-1}	ε_α
311.2	38	6.629	359	532	0.233	0.89
312.2	39	6.996	376	538	0.235	0.90
313.2	40	7.382	394	545	0.238	0.90
314.2	41	7.785	413	552	0.240	0.91
315.2	42	8.208	432	559	0.242	0.91
316.2	43	8.650	452	567	0.245	0.92
317.2	44	9.113	473	574	0.247	0.93
318.2	45	9.597	495	581	0.249	0.93

TABLE A4.   Conversion factors

Length	1 m = 100 cm = 1000 mm
Area	1m^2 = 10, 000 cm^2 = 10^6 mm^2
Volume	1m^3 = 10^6 cm^3 = 10^9 mm^3
Density	1 Mg/m^3 = 10^3 kg/m^{-3} = 1 g/cm^{-3}
Pressure	1 kPa = 10 mb
Heat	1 Joule = 0.2388 cal
Heat flux	1 Watt = 0.8598 kcal/hr
Heat flux density	1 W/m^2 = 0.8598 kcal m^{-2} hr^{-1}
	1 W/m^2 = 1.433 × 10^{-3} cal cm^{-2} min^{-1}
	1 W/m^2 = 2.388 × 10^{-5} cal cm^{-2} s^{-1}

TABLE A5.   Physical constants

Speed of light in vacuum	2.997925 × 10^{8}m/s
Avagadro constant	6.02252 × 10^{23}mol^{-1}
Planck constant	6.6256 × 10^{-34} Js
Gas constant	8.3143 J mol^{-1} C^{-1}
Boltzmann constant	1.38054 × 10^{-23} J C^{-1}
Stefan-Boltzmann constant	5.6697 × 10^{-8} W m^{-2} C^{-4}

Index